Dark Lore

Daily Grail Publishing

Darklore Volume 5

ISBN: 978-0-9807111-4-1

Daily Grail Publishing
Brisbane, Australia
userhelp@dailygrail.com
www.dailygrail.com

Contents

"The most merciful thing in the world, I think, is the inability of the mind to correlate all its contents..."

- H.P. Lovecraft ("The Call of Cthulhu")

If you would like to be notified of future releases of *Darklore*, please send an email to darklore@dailygrail.com. Please be assured your contact details will not be used for any other purpose.

Editor's Introduction

Welcome to Volume 5 of *Darklore*! Regular readers will have noticed that it has been a whole year since the last release. This change in publication interval has been necessitated by a number of factors, not least that it takes quite a while to prepare each volume of *Darklore,* and so – given it is created on a shoestring budget as time allows around 'real life' schedules – the 6-monthly releases became unsustainable. My apologies to those of you who have emailed asking for two, three, and even four releases per year. It just can't be done at this point in time. But we do appreciate your eagerness to see new volumes of *Darklore*!

Some reviews of the previous *Darklore* instalments have criticized the blending of serious, almost academic papers on anomalous topics, with more speculative fare. This led to some thoughtful internal debate as to whether this was an issue that needed to be addressed in future releases. Our conclusion was that it should not – we have made clear in the past that Darklore is not meant as a 'journal' on these topics, as the *Journal of Scientific Exploration* already does a marvellous job of this. What you have in your hands is a compilation of articles on the strange and fantastic that approach these topics from all angles. In this latest release, you'll find articles offering new interpretations based on analysis of primary sources, such as Martin Shough's wonderful analysis of 'the UFO sighting that started it all', and my own piece "How Martin Gardner Bamboozled the Skeptics". Then you'll find well-referenced essays from academics such as David Luke and Robert Schoch, in which they both freely admit that they are speculating, rather than drawing solid conclusions. And then there are explorations of the incredible, and analyses of the careers and influence of two giants of science fiction. In short – a wonderful stroll along some rarely traveled paths, for your enjoyment!

“Dying” (Alex Grey, www.alexgrey.com)

DISEMBODIED EYES

AN INVESTIGATION INTO THE ONTOLOGY OF ENTHEOGENIC ENTITY ENCOUNTERS

by *David Luke*

All that glitters is not gold. Such a maxim might well serve any psychic voyager on a journey into the weirder realms that psychedelics can serve up. After all, out here on the edges there is seldom firm evidence that the beatific or hellish visions beheld whilst chemically neurohacking your wetware have any basis in consensus reality. Indeed these visions are often so extravagantly strange and terrifyingly ineffable that reminding yourself they are *not* real can serve to keep your sanity on a short leash when madness looms. Nevertheless, as John Lilly put it, how does one recognize one's in-sanity from one's out-sanity? And in any case, how would one even begin to try and prove the ontological credibility of the psychedelic experience of visiting some other world or meeting some alien entity? No one has yet put forward a solid method for testing these supposed realities within the domain of science, despite some admirable attempts recently,[1] so all we have left to rely on is anecdote and phenomenology. This story lies somewhere between the two, but it also takes on a new dimension that has urged me to depart momentarily from the fruits of science into the "foamy custard" of folklore and myth, cultural studies and related disciplines.[2] Yet it seemingly has enough semblance of objectivity to warrant a whisper of truth – whatever that may be.

A Brief Glance at the Truly Forbidden...

I'd taken a full DMT dose (~50mg smoked) about forty or fifty times, but always with some trepidation and reverence for its power. True to form, I met a variety of extraordinary entities on these excursions. (As Terence McKenna[3] once said, "You get elves, everybody does.") Sometimes I saw unknown god-like beings, sometimes shape-shifting mischievous imps – but increasingly I kept getting the feeling I was intruding upon a cosmic gathering to which I wasn't invited. Occasionally the effects failed to go any further than an ego-dissolution and a swim through a fractal explosion of pulsing light with the usual wild array of colors. Yet I often felt as though I was being blocked from whatever lurked beyond these multiple geometric dimensions, as well as not being allowed to revisit places to which I had been previously. A couple of times I felt so uninvited and intimidated by the entities I met that I did not wish to return, regardless of my curiosity.

On my last DMT session I was determined to return to the mystic bliss I had once known. I travelled to a secluded beach on the banks of the River Ganges. I prepared myself with an improvised ritual, hoping to gird against whatever lay beyond, and I inhaled a pipe-full of vapors from the foul plastic-tasting resin. Sucked into the space between the pipe and my brain, I found myself breaking through the veil like a gatecrasher into a party of swirling, smiling eyeballs all attached to snake bodies, which were as startled to see me as I was to be there. The whole ordered assortment of eyes and snakes acted as one being. In the brief moment before it reacted to my arrival, I managed to catch a glimpse over what might loosely be described as the 'shoulder' of this strange entity and instantly realized that I had seen something I should not have – a brief glance at the truly forbidden.

Afterwards I could not remember what this *was* exactly, having somehow blocked it out. I only recall that it was a scene that seemed both ineffable and highly illegal for mortal minds. Then the multitudinous eyes of the being before me suddenly and quite deliberately blocked my curious consciousness's further explorations by mesmerizing me with its squirming, rhythmic eyeball hypnosis. I mean, this thing really scared me! It had acted with utter surprise at my being there; and then, alarmed, the ominous numinous proceeded to let me know that I should *not* be there and that I should certainly not be peering into the hallowed space beyond it, which it clearly guarded. I opted against defying this terrifying entity and attempted to remain as passive as possible while it pulsed and gyrated intimidatingly at me for the next ten minutes (though it seemed like an aeon). I finally came out of it all – a bit shell-shocked – and decided that this would be my last DMT experience...for a long time, at least.

Like many of my psychedelic encounters with seemingly discarnate beings, I didn't know quite what to make of this experience, which had rocked me to the core. Yet some time later, after a few years had passed, I began to piece it together with some other visionary fragments. In a dream once, quite naively, I had a mind-blowing encounter with Azrael – the Islamic angel of death. The angel told me its name (which I hadn't heard before), but unfortunately it never showed itself. Among Muslims the archangel Azrael is considered to have ten thousand eyes and it is the holy psychopomp who ushers souls into the realm of the dead. A similar character, Azrail – the god of death – belongs to the Huasa people of western Africa. I also stumbled across Ezekial's vision of the cherubim guarding a throne in the bible (Ezekial 10:12). They, too, were covered in a multitude of eyes, all over their hands, backs, wings, etc., much like the multi-eyed beasts guarding the throne of God in heaven mentioned in Revelations (4:6). These descriptions struck a cord of recognition, although the being I had met on DMT had not seemed quite so angelic.

It wasn't until several years later that I made a surprising discovery, when I accidentally came across a reference in Stephan Beyer's *The Cult of Tara: Magic and Ritual in Tibet* to an ancient deity by the name of *Za* (or *gza'*), who is known to appear with half the body of a snake, no less, and is covered in a thousand eyes. Interestingly, like the cherubim guarding the throne, the Tibetan Za functions as a "protector of the law" and is a guardian deity belonging to a class of demongods called *Lu* or *Lhamayin* (associated with the Indian nāgas), who appear with snake bodies. In Beyer's words: "These *lu* are undisputedly the spirits of the [underworld], found in those places where their realm impinges upon ours, such as in springs, wells, and rivers..." This struck an even greater chord when I realized that on the last occasion I had smoked DMT, I was on the banks of the River Ganges near the Tibetan border, which in retrospect would seem like a sure way to meet this Tibetan deity.

The idea that I had been interloping into the sacred realm of the dead, the underworld, and was blocked by a powerful guardian spirit sat well with my experience, which had me wishing I hadn't turned up unexpected and obviously not on the guest list. Knowing I shouldn't be there, I clearly recall spending the duration of the trip trying to keep my tryptamined mind inconspicuous. And so I focused on the mesmeric rhythmic eyes and nothing more, realizing that I'd stolen a glance at some holy grail when I had burst through the veil. The entity responded quickly and I couldn't have been more compelled to not mess with it. This feeling was further corroborated

The "ancient" protector Za. From an iconographic sketch by Tendzin yongdü (Beyer 1978).

by Beyer who wrote that, in relation to Za and the other fierce protector gods, they:

> ...are the powerful deities who symbolize currents of cosmic force to be tampered with only at one's peril. They constitute the monastic cult [of the Nyingma yogin – the oldest Tibetan sect] because they are best left to the ritual experts. It is not that their cult is particularly secret, just as there is nothing esoteric about the workings of a television set; but in both instances the forces involved are too potent to be played with by a layman, and in both instances the same warning applies.

The same sentiment was echoed by the noted scholar of Tibetan demons, René Nebesky-Wojkowitz,[4] who offered that the Nyingmapa consider the planetary god Za (Rahu) to belong in the highest trinity of deities and that he "guards the religious teachings, and his thousand eyes watch the happenings in the three worlds." Worryingly, Nebesky-Wojkowitz indicates that the elaborate propitiatory cake (*gtor ma*) made to honor Rahu (Za) is constructed of a large red serpentine pyramid dotted geometrically with numerous eyes and bearing stakes "arranged around the base of the 'gtor ma' on which dough effigies of men and animals have been impaled as offerings." (Strangely, this eyed-pyramid bears some resemblance to the be-tentacled pyramidal monster of Robert Anton Wilson and Robert Shea's *Illuminatus! Trilogy,* the Leviathan). Beyer even submits that a lama led him to believe that Nebesky-Wojkowitz died accidentally before his time because of his careless interest in these fierce protector deities.

Gtor ma
(Nebesky-Wojkowitz 1956)

Reading Beyer's account made me feel particularly alarmed that there had been some objective reality to my encounter and that, seemingly, I had actually run into this Tibetan underworld guardian. But these few coincidences barely constitute enough to convince most folks of the objective reality of DMT entities or Tibetan deities. Nor should they, particularly those folks like James Kent[5] who argue that these entities are merely the imaginary output of our neurochemical meddlings. Others have suggested that these entities cannot be considered either real or fictitious but are better thought of as just a part of ourselves.[6] It might have ended there but, soon after, I discovered that my "unique" experience was not so unique after all. And this discovery threatened to bolster the tentative argument that our particular DMT entity, who we could call Za, might have some objective reality – and then so too might all those other beings we encounter along the way to Chapel Perilous, be they mischievous dwarves, machine elves, ancient gods, or praying mantis aliens.

Snake Eyes

Only a few days after reading about Za, I chanced across an article[7] in which the following account appeared:

> I noticed what seemed to be an opening into a large space, like looking through a cave opening to a starry sky. As I approached this I saw that resting in the opening was a large creature, with many arms, somewhat *like an octopus, and all over the arms were eyes*, mostly closed, as if the creature were asleep or slumbering. As I approached it the eyes opened, and it/they became aware of me. It did not seem especially well-disposed towards me, as if it did not wish to be bothered by a mere human, and I had the impression I wasn't going to get past it, so I did not try. [*Emphasis added.*]

That this creature was also quite intimidating and appeared to be guarding the way to something beyond matched my own experience; but it doesn't end there. I was conducting a web survey of paranormal psychedelic experiences at the time,[8] and found that one of my respondents also had a similar experience, but with psilocybin rather than DMT:

> I was convinced I was [dying], I saw another dimension, one filled with eyes in a fibonacci vortex/dome… I've explained this to so many people and regardless of how many things I see, be it in art or biblical references, they all say I'm nuts.

Encouraged by finding these chance reports, I began searching through psychedelic journals and on the Internet for similar stories and found a few more corresponding accounts. This first one occurred with psilocybin-containing mushrooms and appeared in *The Entheogen Review*:

> I began seeing a peculiar phenomenon during low dose mushroom sessions: a pattern of threatening eyeballs. I intuited that the mushroom was trying to scare me, and I marveled at the workings of the mind, feeling humored rather than frightened. ...In spite of my scientifically-orientated worldview, I was being visited by a spirit which seemed to be anticipating a deeper encounter.
>
> ...I took about five grams... This is when I felt the strange spirit enter me: the many-eyed apparition that had already been haunting my consciousness. The difference was that this time the "creature" seemed to be inside of me... I instinctively began questioning its intentions – who was it, what did it want, and was it a demon? I received no answer, and so, not being certain it belonged in my head, [I] forcefully commanded it to leave, which it apparently did...I had the creepy feeling that I was either going crazy or was infected with a spooky denizen of hyperspace... Perhaps, like an insect under a magnifying lens, I have difficulty fathoming this mysterious being of a thousand eyes. Interestingly enough, one of my companions later commented that at one point he perceived my forehead to be covered with eyes.

This next one, posted to Erowid.org,[9] occurred on LSD:

> Countless numbers of eyeballs were looking at me. They were the most evil things I have ever seen. They were all on these snakelike bodies that were weaving back and forth. I reopened my eyes and saw the eyes and the worms all over me and on the ground.

Although I only found these three isolated reports on LSD and psilocybin, I found numerous DMT reports that mentioned eyeball-riddled snake entities in variously weird or disturbing sequences. I needn't quote them all, as this last one offers some kind of "radical empirical"[10] mystical triangulation of my own experience and a tentative interpretation of it:

I remember the veil, like rubber, or the surface of jelly stretched in front of me… I leaned forward to touch the surface of the membrane and then what happened next I swear nearly killed me from its sheer bizzarity… A creature emerged. It was not a happy, smiley elf… It had [innumerable] tentacles, like a cross between some weird octopus or jellyfish…and the EYES! OH MY GOD THE EYES!!!

I froze on the spot thinking shit that's it. I've gone and done it now. I'm fucking toast. I never believed. I should have believed. And now. Now I am at the mercy of [something] much, much, bigger and complex, and clever and definitely [more] malevolent than myself.

I asked it [its] name. I wish I had not asked. [Its] voice utterly destroyed me. It was like being caught in a storm of [psychic] noise – a whirlwind of deadly electrical shrapnel…With its innumerable eyes, It gazed at me steady and extended a tendril. At the same moment it fired a beam of light directly between and above my eyes. The alien laser was pinkish-green. It hurt. I begged it to stop. I whimpered. please stop. you're hurting me. I'm fragile. Please be careful – I am sentient and mean you no harm…

It seemed to consider this; the laser was withdrawn but the tendrils (there were more now) still held me in place. I was trying to make out details of its shape or structure but the closer I looked, the more it slipped away from me. It seemed to tell me in some weird non-verbal fashion not to struggle and to stop making noise with my eyes. I took this to mean 'be calm. do not struggle. clear your head. See but don't look'.

Then it became a little clearer. It seemed to be cloaked in some way – some sort of organic hood and covering was wrapped around

> it – some sort of armour or protection. The tentacles had no substance as we know it and the eyes were the most awe-inspiring/terrifying thing I have ever beheld. They defied counting. They defied reason. The whole thing was [too] much and I felt myself losing my mind.
>
> I...JUST...LOST...IT...goooooooooonnnnnne.[11]

I guess this account really did it for me. There seemed to be at least a degree of objective reality to all these reports (including mine), because they had historical precedent, shared experience, and – most importantly – some apparent meaning. On a level playing field of explanation, where all theoretical perspectives hold equally convincing – or perhaps equally unconvincing – positions, the notion of *meaning* can provide the greatest intuitive appeal to one's understanding. For instance, a physiological or neurotheological explanation might suggest that the highly similar visions are due to similar neurochemical reactions; but this will be seen by some to devalue the complexity and cultural significance of the experience, and it also extends itself much further than the current explanatory power of neuroscience. Alternatively, a parapsychological explanation might suggest that these similar visions all belong to a particular morphogenetic field (a field of consciousness that contains imprints of past experiences which can be accessed by others) activated by chemically-induced near-death-type experiences. Yet there is little understanding of or evidence for morphogenetic fields of this kind, even if they may be possible in principle.[12] Any number of other theories might be put forward. But with all such explanations appearing as equally uncompelling, the possibility of this entity somehow being real as an independently sentient discarnate being – whatever that may be – has comparable explanatory power. However, beyond other ontological speculations this level of explanation – an acceptance of the experience at face

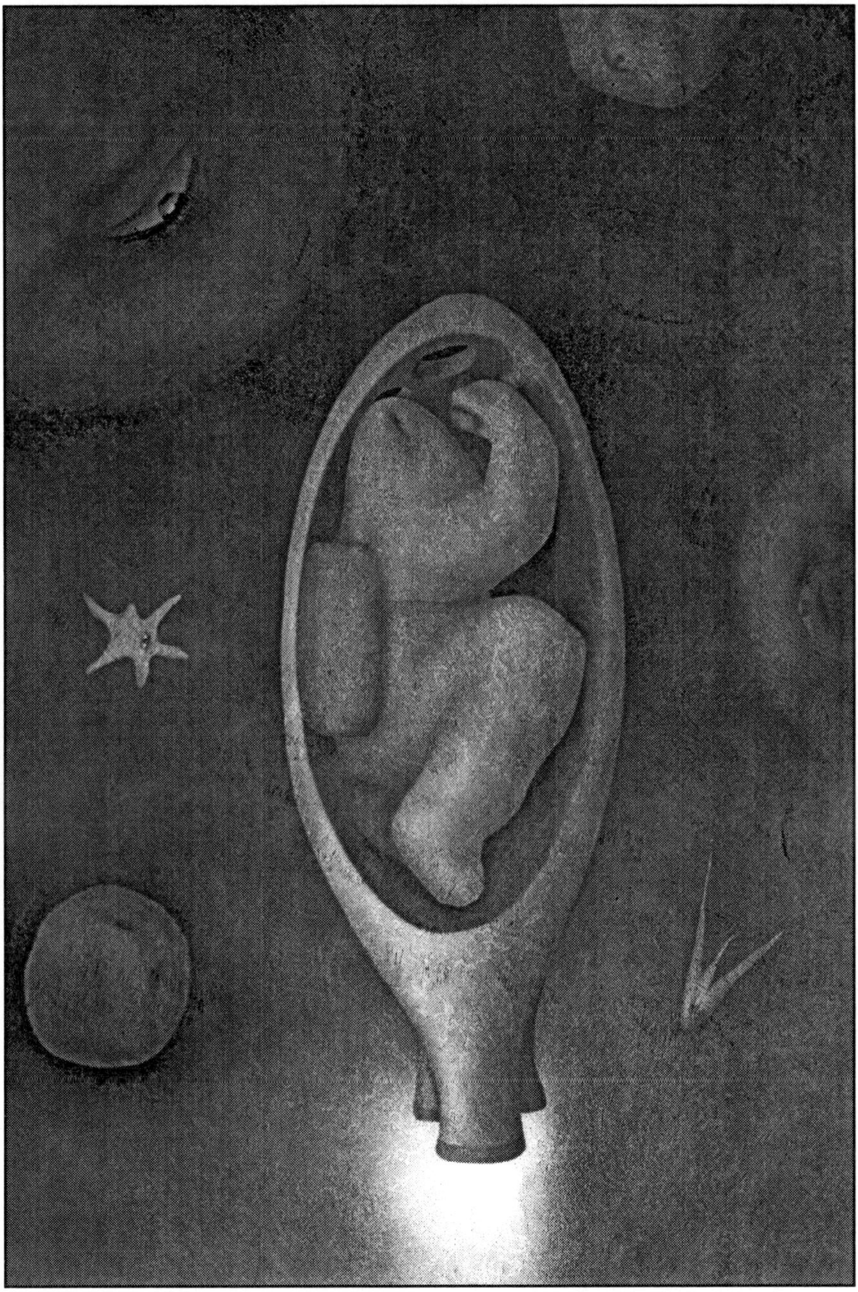

"About Space Flights" (Alex Andreev, www.alexandreev.com)

value – also has esoteric and cultural *meaning* because it fits with a mystical understanding of the universe in which the existence of supernatural beings is accepted.

Sentient Entities

That said, I have little problem, then, assuming that entities – be they dream angels, DMT encounters, or mythical beings – have at least the *possibility* of independent sentience or some kind of objective reality, because I ultimately don't confine myself to any one ontological perspective. So, as clearly as I can make sense of it, it seems that smoking DMT can lead temporarily to some kind of death realm – an idea championed by Rick Strassman[13] and supported by shamanic concepts of ayahuasca states – and in such a place the traveler might encounter one of the (archetypal) guardian deities of the underworld. One such guardian is the angel of death, who appears with thousands of eyes, much like Alex Grey's painting called *Dying.*[14] Yet it seems that sometimes this multi-eyed being also assumes the tentacled or snake-bodied appearance of Za. And like a guardian of the underworld no doubt should be, this being is not to be trifled with. It holds those who encounter it in the grip of utter fear, compelled to obey its hypnotic glare – to just "see but don't look" – because it seemingly guards the sacred way on after death.

On reflection, my encounters with both Za and Azrael have resonance with each other and possibly represent the same psychic atavism or Jungian archetype (albeit an archetype that may have independent sentience), which may become activated by tryptamines such as DMT, or by dreaming or other altered states. This entity is the archetype of the guardian of the realm of death and the doorway to occult knowledge. In considering this, I was lucky enough to find a book by two occultists[15] who offer an argument that the Islamic Azrael, the angel of death, is synonymous with the Hebrew Azazel, the fallen angel of light and the serpent of the Tree of Knowledge (who, as the Promethean prototype, stole the Gnostic fire from God and gave it to man – in much the same way that psychedelics can). They also associate the Persian fallen angel Azza, or Shemyaza with the Luciferian Azazel, who in similar Promethean style swapped the name of God for sexual favors with the mortal Ishtahar, thereby making her immortal.

Jackson and Howard likewise associate Azazel, the great watcher, with the Persian dragon serpent Azhadaha, the black serpent of light and leader of the Inri, the fallen angels known – appropriately – as the watchers. Interestingly, they link the etymology of the common root *az* with the Hebrew letters *ayin* (or *ain* in Arabic) meaning eye, and *zayin* (*zain* in Arabic) meaning sword, which represent the all-seeing eye, and the flaming sword of initiation (the guardian of the Garden of Eden in biblical and cabalistic tradition). Jackson and Howard suggest that, "The secret significance of the Zayin Sword is typified by Azazel as Master of Metals and Lord of the Forge"

Azazel appearing with numerous eyes (from *The Sandman*, Gaiman et al)

because smithcraft and fireworking were the crafts first taught to humans by the watchers, much like the myth of Prometheus. They note that:

> The Hebrew letter-form of Zayin, ז, the sword blade, is the supracosmic fire that, like a shining lightning flash or thunderbolt, "cuts" through the veil of material nescience.

Assembling all these links, it didn't take a huge cognitive leap to also associate the Tibetan eyeballed serpent of my DMT encounter, Za, with these anarchic archangels of other cultures. Without making any great claims to the exclusive resemblance of any of these myths to each other – for these legends have both similarities and differences – further comparisons to Za and Azrael from elsewhere can also be made, such as the Persian Zahhāk, also known in Iranian mythology as Aži Dahāka the serpent or dragon, who was struck down by the divine Frēdōn and snakes issued forth from the wounds.[16] Like Prometheus he was condemned to be chained to the side of a mountain for eternity. The likely etymological link here between the interchanged ayin (a) and zayin (z) of *za* and *az* is itself compelling, especially in the case of the Zahhāk/Aži Dahāka, but the myth story of Za himself has further resonance with the other fallen archangel and Promethean myths.

In Tibetan mythology, Za (known as Rahu in the Indian tradition) features in the *Dri Med Zhel Phreng* version of the Buddhist "churning of the oceans" story about the origins of the original entheogenic ambrosia *par excellence*, amrita, or soma.[17] Having been left in charge of the Buddha's newly made water of life (the amrita) before its supposed dissemination to humanity, Vajrapani (associated with the great soma-fiend Indra) carelessly left the sacred amrita unguarded and returned to find the demon Za, the Lhamayin, had drunk it. In further offence to the gods, Za urinated what remained of the processed amrita back into the vessel. As penance, Vajrapani was made to drink what had now become poisonous, permanently turning blue as a consequence. The similarities here between the methods of enjoying amrita and psychedelic *Amanita* mushrooms have not gone unnoticed, and furthermore the link here between the psychedelic and the Promethean features of the myth is clear.

As just punishment, Vajrapani finally caught up with Za, wounded him many times, and then sliced him in two with his vajra, the lightning bolt. But because Za had drunk the amrita, the water of life, he survived; "amrita" translates from Sanskrit as "deathlessness," and it seems appropriate that this guardian of the underworld himself should become "deathless." As further punishment, the Buddha replaced Za's severed legs with the tail of a serpent or dragon (much like the Iranian Aži Dahāka above) and fixed eyeballs upon his numerous wounds, giving him his unique appearance.

It's here that I saw a further transcultural myth emerging with the legend of the Greek Lamia, the serpentine daimon and prophetess. The Lamia is somewhat similar in character to the Lhamayin, the class of Tibetan serpent spirits to which Za belongs. However there is some contention, not least from the Tibetan scholar, psychedelicist, and etymologist, Mike Crowley,[18] that the Tibetan language has no roots in Middle-Eastern and Mediterranean languages because it is uniquely related to Mongolian. Nevertheless, in the same vein with which Robert Graves (who tipped off Wasson to entheogenic mushrooms) makes more poetic than precise associations between cultural myths, there is a resonance between the legend of Za – the Tibetan serpentine Lhamayin – and the Greek serpentine Lamia, whom we may also associate with Python, the serpentine prophetess of Delphi.

Accordingly, Python was responsible for maintaining the secret of prophecy and the wisdom of the underworld (similarly to Za), was struck down by the sun god Apollo, heralding what Graves describes as the usurpation of the goddess for the rights over divinatory power, and henceforth recasting Python in the role of demon. Something similar also resounds in the Greek myth of the Medusa and Perseus, and perhaps with the Luciferian Norse Loki and the Assyrian-Babylonian Zu (or Azu) too – Zu was struck down by a lightning bolt for stealing the tablets of destiny from Tiamat the dragon queen (but that's another story). With the dawning of the age of patriarchal theism that occurred two to three thousand years ago, the Promethean-type tale of Python retells the same story of the divine maverick: a chthonic being betwixt this world and the underworld, the all-seeing serpent divinity holding the key to man's enlightenment, who steals that wisdom or shares it with mankind and then becomes re-branded as a demon, a fallen angel, a trickster or a deceiver, much like Za, Azazel, and the rest. The Aryan demon Rahu (Za) had once been a Dravidian god and it's clear that an old culture's gods often become a dominating culture's demons, and the archaic tools with which the old culture accessed their divine, be they psychedelic or otherwise, become heretical.

Subsequently, the old chthonic sacramentals, such as amrita, or henbane – called "pythonian" by the ancient Greeks in honor of Python[19] – fell out of grace as easily as Lucifer fell from heaven, or Adam and Eve fell from the Garden of Eden. But like poor old Frank Olsen,[20] did they fall, or were they pushed? The identity of amrita was completely lost, and remains a matter of debate. Although few soma hunters have proposed tryptamines as the culprit – save perhaps McKenna,[21] who championed psilocybin-containing champignons – what the Tibetan lama Chögyam Trungpa says about it fits happily with the various tryptamine visions mentioned above:

> ...amrita is the principle of intoxicating extreme beliefs, belief in ego, and dissolving the boundary between confusion and sanity so that coemergence can be realized.

Perhaps a report of a multi-eyeballed Za-like entity being induced by *Amanita muscaria* might say something more for the favored identity of amrita; and yet, even though there's some certainty that the ancients of the East never smoked DMT, perhaps any old entheogen will do.

But is there anything that can be found in this wayward meandering through myth and vision that offers a case for the genuine reification of "the other" encountered in psychedelic spaces on the far side of the psyche? Knowing that speculation is the vice of the precise and the virtue of the poetic, I have no doubt that those wearing their left brain today will already have departed company with me somewhere along the line here. As a scientist myself, I have deeply questioned this temporary departure from so-called rational thought. But as an explorer of the weirder realms of the mind, I have also been forced occasionally to leap the fence at the edge of my field of expertise and traverse unknown territory. I don't offer any of this as "fact" beyond the phenomenological, but merely as "possibility" in a psychic landscape as "off the map" as that provided by DMT. Indeed, here be dragons. And yes, beware that among the dragon's treasure, all that glitters is not gold. Yet who can resist occasionally inspecting a few gems in case they are of any real value?

David Luke is Senior Lecturer in Psychology at the University of Greenwich and President of the Parapsychological Association. He has published over 50 scholarly articles on anything from pirate utopias to paranormal phenomena with psychedelics. One day he will write a book. He hosts the *Ecology, Cosmos and Consciousness* lecture series in London and is co-editor of www.brainwaving.com.

This article first appeared in *The Entheogen Review*, Volume XVII: I

"3" (Alex Andreev, www.alexandreev.com)

man of the YEAR MILLION

The impossible futures of H.G. Wells

by *Mike Jay*

The young H.G. Wells seems, with hindsight, to have exploded onto the literary world with a procession of breathtakingly original conceptions – *The Time Machine, The Island of Dr. Moreau, The War of the Worlds* – whose diversity and fertility he would struggle to match over the long career ahead of him. But what is rarely appreciated today is that these early works are not a diverse series of imaginative confections, but a sustained polemic on a single theme. That theme is the social ramifications of evolutionary biology, which remained Wells' core concern until the painful overthrow of his theories in 1896, a process documented between the lines of *The Island of Dr.Moreau*.

In trying to use contemporary Darwinian narratives of evolution and degeneration to model the future of humanity, Wells set himself up for bitter failure, but in the process he succeeded resoundingly in fashioning a series of enduring myths of the modern age. Just as the classical myths of metamorphosis, set in the distant past, explored the boundaries between humanity and nature, and our relationship with the animals and the gods, so Wells' stories of evolutionary metamorphoses explore the contemporary questions of what we may become in the future: how the limits of our species might be redrawn, how far our dominance over nature can extend, and whether technology will remain our slave or become our master.

These were questions that had got under Wells' skin in 1884, when he had spent a year at the Normal School of Science in South Kensington under the tutelage and spell of T.H.Huxley. Huxley, like Wells, was a man of humble origin, and his account of evolution was rich with the exhilarating sense that everything must change, and that the history of the world was like a game of snakes and ladders, where those who appeared to be winning were also the ones with furthest to fall. Ten years later, a struggling journalist who had been invalided out of the teaching profession, Wells reached back to Huxley's account of evolution for the big idea that would make his name. It first took the form of an essay, published under the title *Man of the Year Million*, which opened: "As Ruskin has said somewhere, *a propos* of Darwin, it is not what man has been, but what he will be, that should interest us". Man, Wells speculated, was even now in the process of evolving into something unrecognisable. We were losing our ape-like traits – hair, large jaws and ears – and becoming smoother, with "a larger brain and a slighter body". Wells moved incrementally to a haunting image of the man of the far future: floating in a marble pool filled with

nutritive liquid, encased perhaps in a vast crystal dome, by this stage no more than a huge, sensitive hand controlled by a vast brain, the rest of his body shrivelled to a "dangling, degraded pendant". The future man, it seemed, would strike us today as a grotesque alien – a theme to which Wells would return.

But Wells was also aware that, even if the human species were to survive into the far future unchallenged, there were other forces that might destroy it from within. He was particularly under the thrall of the eminent biologist E.Ray Lankester, whose work had highlighted a gloomy sub-plot in the Darwinian story. In his journalism Wells reproduced Lankester's demonstrations that many species appeared to have evolved not onwards and upwards, but into retrograde and degenerate forms. The barnacle, for example, had once been a complex crustacean like a crab, but had lost its powers of locomotion and become no more than a mollusc; the ascidians, or sea-squirts, have highly-developed organs, including eyes almost as sophisticated as those of vertebrates, but have regressed to become shapeless, floating bags that could almost be mistaken for vegetables. Might civilised man, his animal part already atrophying with disuse, be headed for a similar fate?

The Time Machine absorbed all the themes that had been nagging at Wells since his teenage discovery of science, and harnessed them to a narrative on a scale never previously conceived – where all life, up to and including the death of the solar system, was contained within its frame. In the Time Traveller's first stop, above ground live the Eloi, fragile and passive creatures who have begun the journey outlined in *Man of the Year Million*, their animal features receding and their digestive systems no longer able to assimilate meat. Their nemesis and predators, the Morlocks, have made the opposite transit: they are the Coming Beast that Wells had previously proposed alongside the Coming Man, a subterranean throwback to our simian ancestry but also the repository of the industrial and mechanical culture of humanity that the Eloi have lost.

Here were Darwin and Lankester's theories brought viscerally to life for the mass of the reading public who had never been exposed to them in their scientific form. But they were also harnessed to an explicit social warning, equally resonant and as familiar today, that if inequalities of class are allowed to persist and deepen, they may eventually harden into this type of grotesque subspeciation. The Eloi would have been specifically recognisable to Wells' Victorian readers as a satire on the Bloomsbury and Yellow Book set, the art-for-art's-sake aesthetes lazing in the sunny uplands of society while the brutalised workers toiled beneath them. And for Wells himself, the resonance of the Morlocks was a deeply personal one: his mother, a servant at the country estate of Up Park, had spent many years literally working underground. The Time Traveller's journey was fantasy, but it was not escapism: it was a future all too recognisably birthed by the present.

As the Time Traveller ventures onwards to the far future, he finds the implacable processes of degeneration and extinction advancing towards their endgame. The Eloi, in a shocking scene from the serial version cut from the finished novel, are encountered in the twilight of their journey, where they have regressed to scurrying vermin, their human origins still dimly and horribly discernible. After their extinction, giant crabs and sea-slimes claim their dominion over the dying red planet. The calculations of the physicist Lord Kelvin, which had demonstrated that the sun's furnace must burn itself out within a few million years, meant that the deep past so recently revealed would not be granted to the future. However ingenious man's efforts, we were ultimately doomed. The best response available to us was the stoic resignation of the Time Traveller: "If that is so, it remains for us to live as though it were not so".

It was a vision that cut across the prevailing Victorian faith in the forces of social progress, and challenged the complacent view that evolution meant a smooth ascent onwards and upwards towards perfection. Were we really fated to be no more than pillion passengers on a random evolutionary journey, or might we find a way to steer towards a future of our own choosing? How might we master the forces of evolutionary change that were operating on and around us, and shape them to our will? These were the questions that were to be addressed in, around and between the lines of his next major work, *The Island of Dr. Moreau*, Wells' struggle with the defining question of whether and how far the future can be brought under humanity's control.

Plastic Interventions

Wells marked out his early intentions with an essay entitled *The Limits of Individual Plasticity*, in which he questioned how far the humanity of the future would remain bound by our biological inheritance. Could not bodies already be restructured by the newly-developed techniques of 'plastic' surgery and tissue grafting, and minds be reprogrammed by techniques such as hypnosis? "The birth tendencies are only one set of factors" that determine our future. Might new discoveries reveal that we are far more 'plastic', malleable creatures than we have thus far imagined? And how can we know until we have explored these discoveries to their limits?

The character of Dr. Moreau, and his experiments to transform beasts into men, had already begun to take shape in Wells' mind. Wells himself had a good knowledge of anatomy: he had taught it in his previous incarnation, and even written text books on how to dissect rabbits and dogfish. Like Moreau, he knew what it was to have blood on his hands; he was engaged in the current and passionate anti-vivisection debate, which turned for him on the question of how pain was perceived by animals with nervous systems less sophisticated than man's. In another essay, he speculated that pain had arisen as a crude evolutionary adaptation, and the future might see it designed out of the human condition altogether.

But there was another current debate that went to the heart of his questions, and of evolutionary theory itself: could the kind of 'plastic' interventions that Wells had in mind be patched into the hereditary process and transmitted to succeeding generations? To many scientists it seemed that, to some extent at least, this must be possible. Darwin himself had been less than specific about the mechanism by which biological inheritance was actually passed on, but it was well known that alcoholics and syphilitics, for example, risked transmitting their conditions to their progeny. Yet a German biologist named August Weismann had recently made a dramatic intervention, claiming to have shown that acquired characteristics could not be inherited, and that the information passed from parent to offspring was contained in a 'germ-plasm' that passed through the generations unaltered by the life that the parent had led.

Weismann's theory would eventually be recognised as the first statement of what we now know as genetics, but there were many in 1894 who found it implausible, Wells among them. Weismann had no certain answer as to how change, variation and mutation might interrupt his closed system; for Wells, flushed with ideas about how to make natural selection redundant, this uncertainty seemed enough to dismiss his depressing theory. In a science feature in the *Saturday Review* he ridiculed the idea that the mature adult might be perfectly formed in microscosm in the egg, immune to all subsequent influences. But Wells' engagement with this scientific debate would continue in tandem with the writing of Moreau, and the story's eventual resolution would reflect a profound change of heart.

But the book was to be about more than plastic surgery and biology. The society of surgically-altered beast people, and Moreau's attempts to encourage their human traits and repress their bestial ones, were also Wells' entry-point into the contested question of the implications of Darwin's theory for human society, its laws, its ethics and its political organisation.

Darwinian Politics

Although Darwin had publicly proclaimed himself neutral, both his Malthusian logic and several of his private pronouncements supported the thinkers close to him, such as Herbert Spencer, who claimed his mantle in the name of *laissez-faire* economics, imperialism and the dominance of the civilised races. The term 'survival of the fittest' had been coined by Spencer, and Darwin had later admitted preferring it to his own 'natural selection'. For 'social Darwininsts' like Spencer, the elimination of the poor and technologically backward might seem harsh to human sensibilities, but had to be seen in the context of the "decrees of a large, far-seeing Providence" that was using the laws Darwin had revealed to progress to a better world. "The study of nature", as Wells would have Dr. Moreau say, "makes man at last as remorseless as nature".

But there were others who claimed that Darwinism had entirely different political implications. Karl Marx had wished to dedicate *Das Kapital* to Darwin, on the grounds that it showed "the basis in natural science for the class struggle in history". Darwin's co-discoverer of evolution, Alfred Russel Wallace, believed that his theory demonstrated that a world-socialist state was the only form of government that would allow man's higher nature to flourish. The anarchist prince Peter Kropotkin, lecturing in London in 1893, argued that Darwin had shown society's natural form of organisation to be the self-organising and mutual cooperative, and that the Darwinian selection of political systems would eventually transcend the already atrophying nation-state. All agreed that Darwin's theories had set these age-old questions of politics on a scientific footing for the first time, and that science must prevail in pointing the way forward; but they disagreed profoundly about how to read the signposts.

Wells had already written about the paradoxes of competition and cooperation: how, from the early stages of life such as sponges and coral, communal structures had delivered paradoxical advantages in the struggle for existence. But, just as Weismann's experiments were questioning the limits of plasticity, the foundations of the 'social Darwinist' debate were being shaken by Wells' old mentor, T. H. Huxley. Huxley had previously been associated with a harsh and imperialist interpretation of Darwinism, stressing the 'gladiatorial' nature of natural selection; but in his lecture on Evolution and Ethics at Oxford in 1893 he had dismissed the entire notion that Darwin's theories might have anything to teach us about how to organise society. The state, he insisted, should not be an extension of nature but a corrective to her tyranny. Civilised, or ethical man, was formed by the pursuit of justice, a concept entirely alien to nature. Early civilisations had developed passive philosophies, Stoicism or Buddhism, to accept this truth; but modern man was now empowered to fight back against nature, and to make life humane and civilized as never before.

As Wells worked through three drafts of Moreau, he found himself paring back his early optimism about plasticity and technologically-enhanced evolution, and becoming more sceptical about Dr. Moreau's crusade to borrow nature's tools, pain and fear, to discipline his beast-people into their surgically created humanity. Moreau became an ever more tragic figure, resigned to the realisation that his noble experiments on the beast-men were nothing but pointless cruelty that left his hybrid creations 'stumbling in the shackles of humanity' and reverting inexorably to the mindless beasts that they had been before his radical attempts to improve them. This conclusion had also developed into an unmistakable critique of a Christian God who had created humanity in a crucible of pain and sin, and proceeded to use the threat of everlasting torment to keep it obedient.

Moreau laments the futility of his efforts to buck nature, efforts which are reversed by no greater power than the ineluctable passing of time:

> Just after I make them, they seem to be indisputable human beings. It's afterwards as I observe them that the persuasion fades. First one animal trait, then another, creeps to the surface and stares at me…

The book finally emerged with a vision even bleaker than that of *The Time Machine.* The future of Eloi and Morlock, dark though it had been, had at least had a logic: it had emerged by a chain of cause and effect from human actions passed down the generations. In the world of *The Time Machine*, humanity might have changed its ways and escaped its destiny. But now, he accepted, it was a vision that had been based on scientific error. The story of Dr. Moreau showed that, however much we might wish it otherwise, "natural selection grips us more firmly than it ever did". If we wish to improve humanity's lot, we must do so in the teeth of nature rather than in harmony with her. "The business of science", Wells concluded grimly, "is not to keep up the courage of men, but to tell the truth".

The End of Evolution

Wells was now, reluctantly, convinced that there was no escape from the *cul-de-sac* of nature and its tyrannical laws; yet it was clear to him that the future was none the less upon us, and needed urgently to be anticipated. His next move would be the *volte-face* that would set him on his mature career – and, by the same token, betray the vision of his youth. By his thirtieth birthday, in September 1896, he had produced not only the scientific essays that would underpin his new strategy for the future, but also his most successful novel yet, *The War of the Worlds*: a story that would encapsulate this revised future in mythic form, and would long outlive the theories that had birthed it.

His change of direction was set out in a long article entitled *Human Evolution: An Artificial Process,* in which he effectively rejected all the assumptions about human evolution and plasticity that had driven *The Time Machine* and *The Island of Dr. Moreau*. He bowed once more to Weismann's conclusion that neither instigating new habits nor abandoning old ones would affect humanity's future evolution. He considered the emerging claim that the human race, in the absence of the selection pressures that had formed us, has probably to all intents and purposes stopped evolving already. He conceded that the huge disparity in scale between our slow biological reproduction and our dizzying technological acceleration made it impossible to know for certain which, if either, would turn out to define the inheritors of our civilisation. But his notion of 'artificial' evolution is, in plain language, education, and the transition from plastic to artificial evolution marks the beginning of the transition from the young Wells, the cosmic iconoclast, to the mature Wells, the educator and statesman. His claim for the future was now that:

> The attainment of an unstable and intransitory perfection only through innumerable generations of suffering and "elimination" is not necessarily the destiny of humanity...in Education lies the possible salvation of mankind from misery and sin...

It was a claim that demolished the foundations of all his major work to date, but it was ammunition for the argument that would drive his thinking for decades to come. In evolutionary terms, he conceded, there was no difference to speak of between Palaeolithic man and his modern-day successors. The difference, though vast, was wholly an artificial one: an accretion of discovery and technology and civilisation that the man of ten thousand years ago, if transported to a modern metropolis, would have no chance of comprehending. Might not the Man of the Future, then, still be able to supersede the man of the present in ways that we are yet unable to imagine? The laws of nature might be immutable, but artificial evolution on this scale could make them irrelevant.

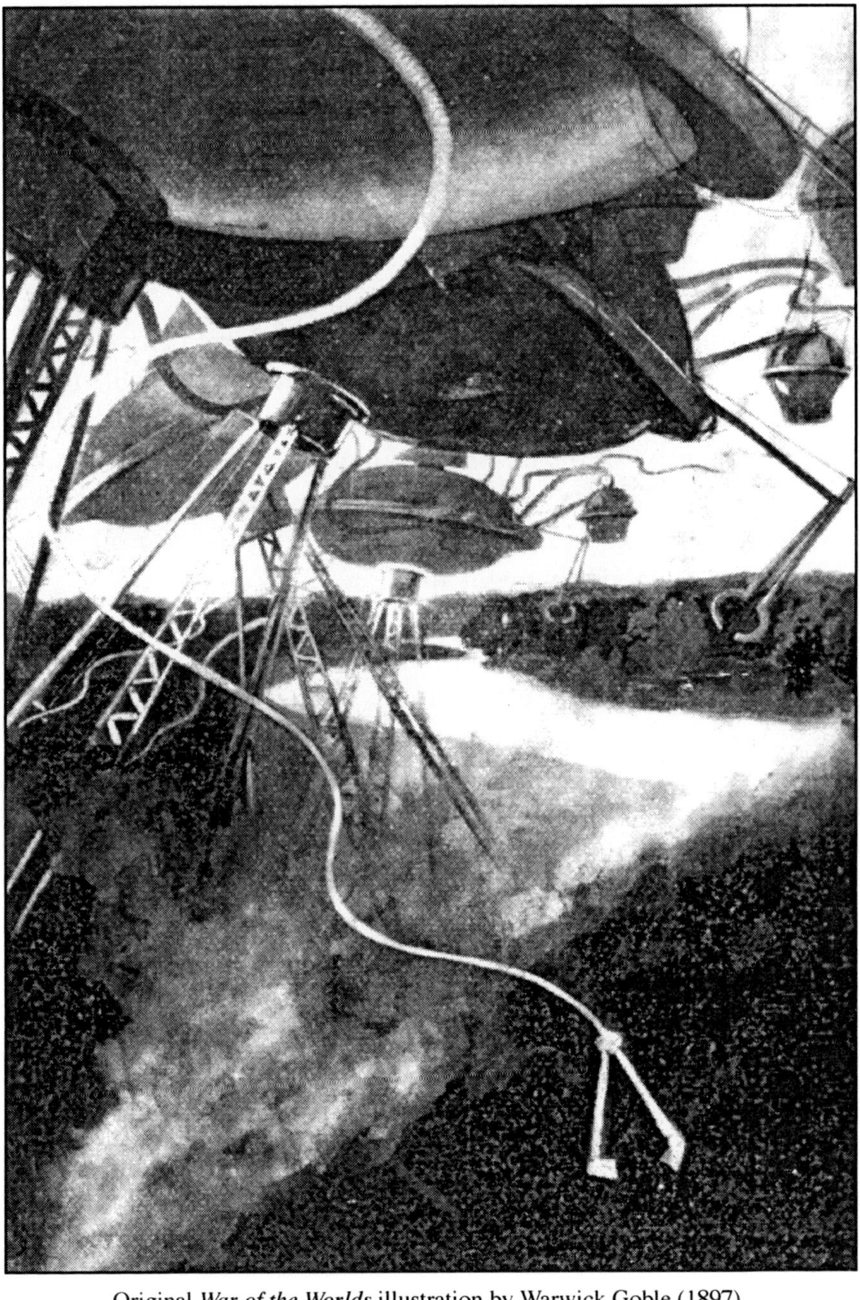

Original *War of the Worlds* illustration by Warwick Goble (1897)

To put flesh on the bones of an artificially-evolved future, Wells began to imagine alien life-forms that might have embarked on such a process in the distant past, and how they might look to us now. He wrote various essays proposing life in other galaxies, or based on silicon chemistry rather than carbon, that look forward to the artificial life simulations of the present day; he paid particularly close attention to another debate equally familiar today, as to whether the desert planet Mars might have had water, and thus life, in its distant past – and indeed whether, as the Astronomer Royal Percival Lowell proposed, it might today be home to desiccated and dying life-forms. He considered the fate of the Tasmanian people, who loomed large in the conscience of the civilised world as the first race deliberately driven to extinction by colonists who countered their spears with modern rifles, and tried to imagine how an alien race of similar technological superiority might appear to us.

This was, of course, an act of the imagination that he had already undertaken, and with which he had begun his journey into the future. The final stage of humanity that he had envisaged in *Man of the Year Million* was as alien to humanity as anything from a distant galaxy: giant brains, the rest of their bodies reduced to vestigial sacs except for a super-sensitive hand to transmit the brain's instructions to the advanced technology that had become their exo-skeleton. There was a pleasing irony in demonstrating the possible scope of artificial evolution by visualising the human race being annihilated by our future selves. Cycling around Woking and along the Thames towards London, Wells began to trace the route of an invasion by Martians who, for all their outlandish tripods and tentacles, were in reality our own great-grandchildren.

When *War of the Worlds* was launched on the public during the Jubilee year of 1897, its impact was sensational, and its message the strongest possible contrast to the national mood of celebration and self-congratulation. It punctured the sense that Britain stood at the summit of civilisation by conjuring all too vividly how brittle such a civilisation really was: how rapidly an unforeseen crisis might shatter it, and how little consolation our traditions or religious faith would have to offer. In visualising the wholesale destruction of modernity, Wells was not mocking the idea of progress, but of complacency. This was iconoclasm with a purpose: to spur his readers on to the remaking of the world, to which Wells would now turn his own energies. For the next fifty years he would seek to drive humanity towards social and political change, towards open societies and world government; but although he would continue to collaborate with scientists like Lankester in other capacities, he would never return to the narratives of biological evolution that had been the ruling paradigm of his greatest work.

Mike Jay's latest book is *High Society: Mind-Altering Drugs in History and Culture*, published in November by Thames & Hudson (UK/Australia) and Inner Traditions (USA). It accompanies the exhibition he is curating at Wellcome Collection in London (Nov 2010 - Feb 2011). His website is http://mikejay.net.

Original Wilton Windmill photo courtesy of Steve Alexander, www.temporarytemples.co.uk [1]

Closing the CIRCLE

Enchantment and disenchantment in the fields of Wiltshire

by *Richard J. Andrews*

Sometime in the very early hours of Monday May 24, 2010 – when I really ought to have been getting some sleep in preparation for the working week ahead rather than getting a social media fix – I followed a link to a photograph of a crop pattern reported the previous Saturday. Named for its immediate proximity to Wilton Windmill in Wiltshire, it was a simple circle: 300 feet in diameter, and formed in five-foot-tall oilseed rape. It was inscribed with twelve radial spokes, from either side of which extended eight short arcs, constructed on circumpolar concentric circles.

Noting its superficial similarity to the disc created at Crabwood Farm in 2002, which was accompanied by a cleverly implemented representation of an archetypal alien face, and located next to a microwave relay station, I immediately suspected that a message could be encoded into the Wilton Windmill formation's design and decided there and then to have a crack at it. The Crabwood disc had encoded a spirally-arranged and highly enigmatic message in binary code, translating into English via ASCII character codes – numerical representations of keyboard characters used by computers.

> Beware the bearers of FALSE gifts & their BROKEN PROMISES. Much PAIN but still time. Believe. There is GOOD out there. We OPpose DECEPTION. COnduit CLOSING,

The eight arcs attached to each spoke of the Wilton Windmill crop pattern strongly hinted at the possibility of 8-bit ASCII codes (and as others have noted, the arrangement of the code is reminiscent of the arrangement of data in sectors on a computer hard drive), but I needed to figure out which way to read them. Because the inner ring of arcs all extended from the spokes in the same anticlockwise direction, I decided that these probably represented redundant zeros, placed in the first column of each byte of information. Each letter was to be read from the inside of the circle towards the outside. Next I had to decide where the message began. The Wilton Windmill itself provided a useful punctuation point by the side of the field, so I assumed the message should be read from there, in a clockwise direction.

After tabulating the binary digits on the ubiquitously-employed back of an envelope, I typed the supposed 8-bit character codes into a handy online ASCII conversion utility to derive the corresponding keyboard characters. The resulting string of characters read:

e^(hi)pi)1=0

Hidden Identity

It looked like some kind of equation, and when I copied it into Google and hit the search button, Google asked if I had meant 'e^(i)pi)1=0', for which the top result was something called 'Euler's identity', usually written as $e^{i\pi}+1=0$.

Euler's Identity has been rated as one of the most beautiful expressions in mathematics. One aspect of this beauty is the fact that it combines, in highly compact form, five fundamental mathematical constants with three basic arithmetic operations. It has also been said that it "reaches down into the very depths of existence" – an impressive claim.[2] Given that the fields of Wiltshire have been amply exploited in recent years as a canvas for the simultaneous expression of key mathematical principles and aesthetically beautiful geometrical forms, I suppose it was almost inevitable it should turn up there eventually.

Euler's Identity is a special case of Euler's formula, represented graphically by a circle inscribed by a unit radius passing through all possible angles. That this particular formula has to do with the mathematics of the circle and the cyclic oscillation of trigonometric functions makes it a perfect subject for a 'crop circle'. On first seeing the photographs of the crop formation, I had been struck by the subjective impression – perhaps reflecting my personal interpretative bias, that it referenced both the turning wheel of the windmill and the twelve-part division of the zodiacal cycle, the cosmic wheel. Euler's Identity seemed to make perfect sense of those references. On reflection, I felt that this was a very clever and elegant design, in which mathematical and symbolic meanings were fused into a single 'identity'.

My excitement at having decoded the formation,[3] and my admiration for the designer, was somewhat marred however by a number of apparent errors. For a start, the translated message e^(hi) pi)1=0 contains an anomalous 'h'. It is possible that this was an error,

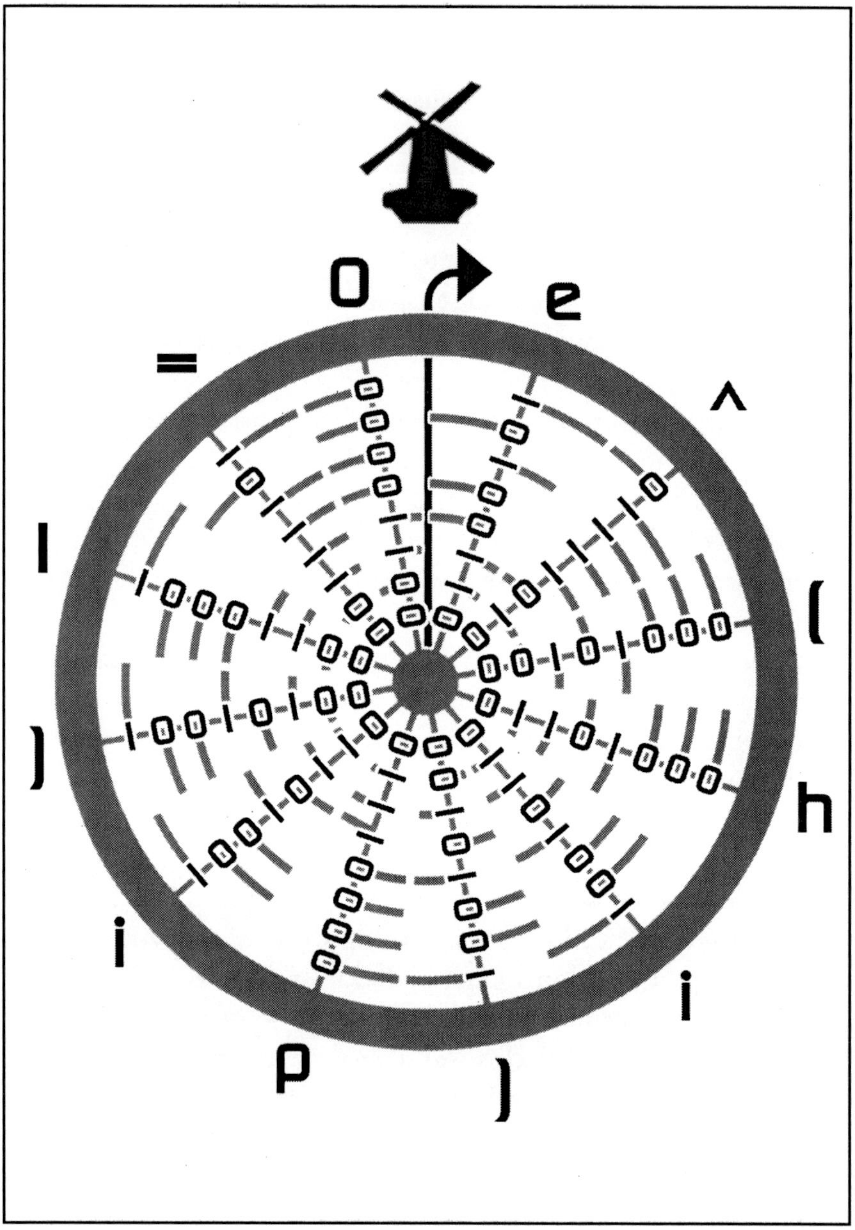

Binary-ASCII encoding of Euler's Identity in Wilton Windmill crop circle[4]

as the ASCII code for this letter is just one 'flipped bit' from that for the opening parenthesis '(' (01101000 as opposed to 00101000), which is needed to pair off with the otherwise unpaired final closing parenthesis. This would give us e^((i)pi)1=0. But there is also a '+' missing, a crucial mathematical component, which if introduced, would bring the number of characters to thirteen: e^((i)pi)+1=0. However, a circle with thirteen equally-spaced spokes would perhaps be harder to create in the field, and less symbolically potent.

Again, we could be looking at a flipped bit. Turning the closing parenthesis ')' (00101001) into a plus sign '+' (00101011) would give us e^(hi)pi+1=0. However, if we correct both of these 'errors', to give e^((i)pi+1=0 we are still left with an unpaired parenthesis.

There are several more accurate renderings of Euler's Identity which do use twelve characters. Both e^(i*pi)+1=0 and e^(i.pi)+1=0 would work. Since pi (π) has its own binary ASCII character code of 11100011 (or 227 in decimal notation) there is another possibility: e^((i)π)+1=0.

Whatever the cause of the 'error', it seems to have a message of its own. Perhaps it is no accident that the 'h', with the adjacent 'i', reads 'hi' – an embedded greeting from the circle-maker?

When someone mentioned on a forum thread, in all innocence, that 'h' was also a constant (the Planck constant), I couldn't help laughing (although sharing my mirth with the world via a three-letter acronym was easier to resist). The out-of-place intrusion of a constant from the world of physics into the abstract mathematics of Euler's Identity makes a most appropriate metaphor for a mislaid 'plank' (or stomping board) – the classic circle-making implement – bringing the 'pi in the sky' notions of non-human circle-makers crashing down to earth. Maybe it was just a Freudian slip, but I wonder if anyone thought to check the flipped bit in the field for a lost stomping board?!

Ambiguous Messages

Despite having taken the trouble to visit a local crop formation in the early 90s, I would not count myself amongst the 'croppies' – those who follow the phenomenon closely. I do, however, take a general interest in things anomalous, and as a result of my experience, I guess I'd be happy with the title of 'honorary croppy'. I'm not sure I would be universally accepted into their number though, since I am of the opinion that crop patterns are made exclusively by human hand – as my flippant interpretation of the flipped bit suggests. The reader can take it as read that the following arguments are based on this premise. I don't intend to justify my stance here, beyond stating that this is the only proven cause of the phenomenon to date.

However, those who claim to make crop patterns have been understandably reluctant to incriminate themselves by providing specific evidence of their involvement. The unsubstantiated claim that all crop patterns are human-made thus leaves ample room for ongoing mystery and intrigue surrounding this beautiful and enigmatic phenomenon, despite a creeping realisation amongst croppies that at least some crop patterns are human-made.

Of the alternative theories regarding crop pattern origins, the most prevalent, since the advent of complex designs, as opposed to simple crop circles, is that they are a medium through which some kind of non-human intelligence is attempting to communicate with humankind. Clearly, the makers of the Wilton Windmill crop pattern are not so intelligent as to avoid making basic mistakes. Errors in crop patterns can be revealing of the fact that their makers are only human after all. The pressure of executing an exacting design in near darkness and silence within a limited time window is bound to lead to the occasional error.

Such errors in crop patterns are probably more frequent than generally realised. They can be hard to identify when the original intentions of the designer are unknown, or if the concepts intended to be conveyed by the design are abstract, ambiguous, or even non-existent, relying to a greater extent on the subjective interpretation of the viewer. However, the inclusion of a highly specific encoded written message is a high risk endeavour. Not only does design complexity increase the probability of error, but any errors will be easier to spot; the decoded message of the Crabwood Farm crop pattern included a number of anomalies – partly capitalised words – which might be best explained as human error.

The 'message' in the case of the Wilton Windmill formation was relatively easy to decode, despite its anomalies, for a human – and it takes one to know one. Multiple human cultural conventions are employed in its design, such as computer code, clockwise rotation and mathematical notation (not to mention it's measured diameter, the suspiciously round figure of 300 ft). The symbolic windmill location also implies familiarity with their function as well as local geographical knowledge.

The fact that the occasional crop formation contains a relatively unambiguous decodable message encourages croppies to attempt to interpret the meaning of every crop pattern, regardless of any

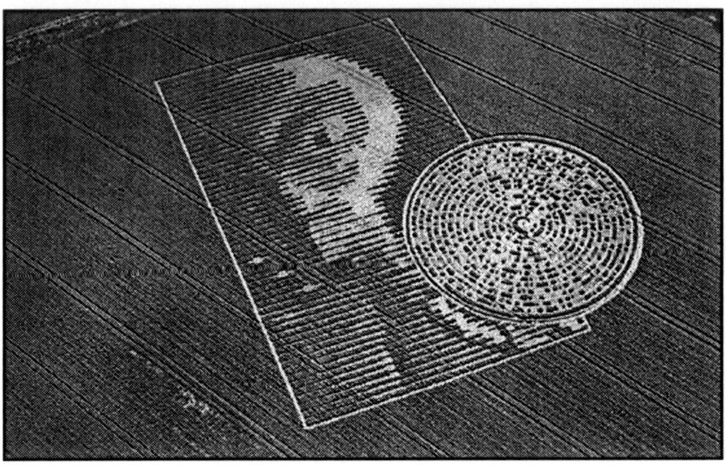

Crabwood 'alien disc' crop circle
(Steve Alexander, www.temporarytemples.co.uk)

intent on the part of the makers to encode a message. Would-be conspiracy theorists might wish to comment on the resemblance of this circumstance to what psychologists call a 'variable ratio reward schedule', a highly-effective behaviour modification technique!

Even with a message as unambiguous as a mathematical formula, there is much scope for subjective interpretation. In this respect, the patterns act like Rorschach tests, allowing their observers to project onto them their own fantasies. I had perhaps naively expected my swift decoding of the Wilton Windmill formation to be the last word in the interpretation of its meaning, but I couldn't have been more wrong. Croppies as a group are very keen to project their own diverse and often far-fetched interpretations onto every detail of a crop pattern.

According to circle-maker Rob Dickinson, the ambiguity of a typical crop pattern is intended to "provoke inquiry and challenge pre-conceived ideas".[5] This stated intention originates from the early years of crop-circle making, but whether or not it is shared by other circle-makers, all too often, the result of the circle-makers' art is merely to perpetuate entrenched belief systems.

Ambiguous Messengers

The modern phenomenon of crop patterns allegedly originated with artists Doug Bower and Dave Chorley, who claimed that their simple crop circles in Hampshire were an attempt to emulate the so-called 'saucer nests' which were a feature of the Warminster UFO flap of the 1960s and early 70s, and which Doug had also come across in Australia. UFO research is notoriously plagued by hoaxes, and many of the hoaxers have a genuine interest in UFOs. Such was the case with Doug and Dave.

Eventually, their efforts attracted the attention of the mainstream media, and physicist Dr Terence Meaden, called in to investigate the phenomenon, came up with the meteorological 'plasma vortex' theory.

Doug and Dave were apparently displeased with this development, and began to create more elaborate designs to confound him, for example adding four smaller circles around a larger one as if to represent the 'feet' of a flying saucer. Meaden, however, simply adapted his theory.[6]

By 1990, 'pictogram' designs with straight lines, and even double-D signatures – the first encoded messages – began to appear, not that anyone could decode them at the time. It seems that the original plan to hoax supposed UFO landing spots had been forgotten, and a new game of cat-and-mouse was afoot. The development of crop pictograms suggested an intelligence behind the designs, assumed by many croppies to be non-human, and offered much scope for symbolic interpretation.

The phenomenon started to occur over a wider area, and Doug and Dave surmised that other teams must have begun to operate, creating for example the double pictogram at Alton Barnes immortalised on a Led Zeppelin album cover. It seems that some croppies, in an act of boundary-crossing typical of the paranormal field, were creating crop patterns themselves – perhaps to disprove the non-human-origin hypothesis, or perhaps in an attempt to establish a dialogue with the still-mysterious makers. Once they had sampled the dark art of circle-making, blurring the distinction between the seeker and the sought for, the culprit and the victim, as servants of the Trickster[7] their motivations became harder to discern, perhaps even to themselves.

The belief that there is more to this phenomenon than teams of young men competing to create the best illegal art under cover of darkness is still prevalent in some circles (pun obviously intended). The croppies' reluctant acknowledgement that some crop patterns are made by human hand, combined with their unshakeable conviction that others have a more mysterious origin, leads them to make a distinction between 'genuine' and 'hoax' formations.

Many croppies can not bring themselves to believe that the hated 'hoaxers' could have anything enlightening to say, so the presence of a 'message', whether real or imagined, has either been taken as evidence of a genuine crop formation, or the work of disinformation agents. They would rather resolve their cognitive dissonance by touting paranoid conspiracy theories, than by admitting to their own folly.

To state that the perpetrators of human-made crop patterns are hoaxers – deliberately attempting to deceive – can be problematic when there is no evidence of their making a false claim. This hair-splitting may be usefully exploited by the deceitful, of course, and who can say that the circle-makers, mixing amongst the croppies at their Wiltshire watering holes and conference venues, are not guilty of introducing false interpretations of their own work? Not that the croppies, willing partners in their own delusion, would need much help in that department!

The concept of projection, in its original psychiatric context, refers to an unconscious defence mechanism whereby what is emotionally unacceptable in the self is rejected and attributed to others. By attributing the role of hoaxer to the circle-makers, might the croppies be rejecting the thought of their own self-deception?

Perhaps it is the self-appointed and phenomenologically-focused expert 'researchers', intent on promoting the concept of non-human origin, who might be justifiably accused of making false claims. As Carl Sagan was fond of saying, "extraordinary claims require extraordinary evidence". Many factors have been cited as evidence of 'genuineness', but science advances by falsifying explicit claims, not by proving vague ones. While a hoax requires only the theoretical existence of a genuine counterpart (such as Bigfoot), a 'genuine' event implies a unique and known origin, yet the mystery of crop patterns relies on the claim that their origin is not known.

Such considerations highlight the questionable pseudoscientific methodology of the 'researchers'. Should they also stand accused of pseudospirituality, of failing to remove the metaphorical plank from their own eye, before attempting to remove the speck from the eye of their brother?

The evolution of the crop patterns from saucer nests to sacred art has been accompanied by a parallel evolution in the definition of the 'genuine' article by croppies, and the expressed motives of the circle-makers. It is the uneasy symbiotic relationship between producer and consumer that has driven this evolution. This feedback loop further blurs the distinction between subject and object, scientist and artist. The swirling vortex of a crop circle is an appropriate image of the dynamic reflexivity at the heart of this mercurial phenomenon.[8]

A Genuine Mystery?

Many of the croppies, bewildered by the various claims and counter claims, now state that what matters is the personal experience of viewing and entering the designs: the awe that they inspire. Arguably the effects of crop patterns are now more interesting than their causes.

This approach to crop patterns neatly transcends the croppies' lack of consensus on the hoax/genuine issue, enabling them to accommodate to the human aspect whilst maintaining the spiritual, by invoking the agency of divine inspiration. It also demonstrates the transcendent nature of experience as neither genuine or false, real or unreal, and helps neutralise the negative feelings projected onto the circle-makers by those still taken in by the hoax mentality.

Given their generally secretive and deceptive nature, the question of the circle-makers motivation is one of the genuine mysteries of this phenomenon. We are all human, and I suspect that the croppies and the circle-makers are equally prone to engage in rationalisation. As circle-maker Jim Schnabel has stated, "If there can be artifice on the way into the mind, there can be artifice on the way out".[9] One interpretation of this statement could be that circle-makers can deceive themselves as well as others with regard to their motivations and actions. Latterly, as we have seen, some circle-makers have adopted a high intellectual stance, promoting their work as crop art, as an invitation to look within ourselves and understand the workings of our own minds, perhaps in the hope of retrospectively shaking off their reputation as lowly pranksters.

Another interpretation is that the circle-makers have come to the realisation that art is magic. Like stage magicians, their work relies upon the fact that people just don't suspect the incredible lengths they go to in order to pull off a trick, to create that all-important 'wow' (or is it 'woo'?) factor. While the art of deception can trigger paranormal or spiritual experience, it is not the source of that experience, but instead a way of distracting the mind, of opening it up to novel possibilities. Like a temporary temple, a religious icon or mandala, a crop pattern merely provides a supporting context, an appropriate setting, where belief in the divine can work regardless of its proven existence. It matters not whence it comes.

Their reflexivity qualifies the circle-makers as 'authors of the impossible', in Jeffrey Kipal's terms.[10] Kripal identifies two stages to becoming an author of the impossible: Realisation – recognition of our participatory role in the paranormal, of our "reading the paranormal writing us"; followed by Authorization – stepping out of the story we have been written into, and "*writing* the paranormal writing us".

The words of Alan Moore are pertinent here:

> It is not the job of the artist to give the audience what the audience *want.* If the audience knew what they needed, then they wouldn't be the audience, they would be the artist. It is the job of artists to give the audience what they *need.*[11]

Richard J. Andrews is a freelance web professional and aspiring author. He has studied environmental psychology at postgraduate level and has a Diploma in Heritage Interpretation. Some of you may know him as Daily Grail admin 'Perceval'.

secrets of the angelic society

The ongoing quest to return to the Golden Age

by ***Philip Coppens***

On December 31, 1986, Andrew Lakey suffered a cocaine overdose. For the first time in twenty years Lakey, then 27, prayed, telling God that if he survived, he would never do drugs again. He did survive, and then – apparently out of nowhere – he began to draw angels. Four years later, he confided what had truly happened. As he showered, hoping to counter the effects of the drug overdose...

> ...I saw seven angels twirling around my feet. Eventually they came together as one angel, its arms wrapped around me. I fell to the bottom of the shower, but I was in another space.

He explained that the angels had told him to take up painting, even though Lakey had never received any training. Since then, Lakey has painted thousands of angels and is renowned for using a three-dimensional technique which has allowed the blind to experience his art – the legendary Ray Charles was amongst his many fans. His first angel painting ever currently hangs inside the Vatican.

Lakey is just one example of a well-known artist who was not merely inspired, but transformed by an intense experience with angels. The angels 'made' Lakey a world-renowned artist, instilling him with an ability to paint which he did not acquire through any human training. However, when you begin to look into the history of art, it soon becomes apparent that many artists throughout the centuries have claimed to have had angelic contacts. In fact, digging even deeper, it becomes apparent that many have claimed to be part of a secret group, each of which have left codes in their paintings to show – to their fellow initiates – that they too are members of the 'Angelic Society'.

Et in Arcadia Ego...

According to the French writer Maurice Barrès, this code is said to be the Latin phrase ET IN ARCADIA EGO ("And in Arcadia I"). The most famous incorporation of this phrase is in *The Arcadian Shepherds*, created by the 17th century French painter Nicolas Poussin. A copy of the painting, with its enigmatic inscription, can also be found on his memorial slab in the church of San Lorenzo in Lucina in Rome, underlining how dear to his heart this painting and phrase was. There has been intense speculation for decades that Poussin was more than just a painter; the 'Sun King' Louis XIV decided to hang this painting in his private bedroom, apparently believing that the painting contained the solution to a mystery. Alas, there is so much intrigue surrounding Louis XIV that it is far too difficult to untangle that conundrum in the space available here. But what is clear, is that in the eyes of Barrès, Poussin's life and career, somehow

Nicolas Poussin's *The Arcadian Shepherds*

had a connection with angels. And Barrès knew what he was talking about, as he himself was angelically inspired.

Born on August 19, 1862, Maurice Barrès was politically active throughout his life, and in 1920 influenced the French government to institute a national feast day for Joan of Arc – a woman who also claimed to have intense communication with angels, including the Archangel Michael. Indeed it was these angelic voices that gave her counsel on how to fight the English during the Hundred Years' War. If ever there was a patron saint for those who were guided by angels, Joan of Arc would be it...and perhaps she is!

Barrès was also friends with Claude Debussy and Victor Hugo, both of whom have been listed as "Grand Masters" of the Priory of Sion. As fictional as this list is – there never was a genuine Priory of Sion, and hence no grand masters – it is interesting to note that a number of the famous personages listed shared an interest in the topic of contact with angels. Barrès was also a very good friend, from childhood, of the occultist Stanislav de Guaita; they both attended the same school in

Nantes around 1880. He introduced Barrès into Martinism, an occult practice very much in vogue at the end of the 19th century, especially in French esoteric circles. The preface for one of the editions of de Guaita's *Au seuil du mystère* (1886) was written by Barrès.

But despite his close friendships with the leading lights of French esoterica, Barrès is seldom perceived as an "initiate". Instead, he is presented simply as an 'admirer' of de Guaita, which greatly misinterprets their bond…as well as Barrès' intelligence. Perhaps the problem is that his role as an initiate was only revealed posthumously, when *Le Mystère en Pleine Lumière* was published in 1926. The book is something of a memoir, in which Barrès weaves in certain parts of his life with the lives of people he held in high regard. It is in this book that he underlines his devotion to Joan of Arc. He stipulates that she and others acted as guides for Barrès on his path of "initiation". It is in this book that he talks about a "mystical brotherhood" whose existence is made discretely visible by the afore-mentioned inscription: "It is required to leave, in a certain manner, in a certain part of our work, a tombstone which contains the famous inscription: ET IN ARCADIA EGO."

Why the need for secrecy? Barrès believed that there were a small number of individuals who could "enter in direct contact with God". Of course, direct experience of the divine is precisely the key differentiator which has set, for two millennia, the Church against the heretics. The Church argues that God can only be experienced indirectly, via the Church hierarchy, whereas the gnostically inclined heretics believed that a personal experience of the divine was possible and should be experienced, for it had tremendous transformational powers. Andrew Lakey is certainly a perfect example of this.

The ensuing battle between both traditions is – literally – history. No wonder therefore that those who experienced such direct contact with the Otherworld were most discrete about it, using clues like obscure Latin phrases in their paintings. And though there are various means of experiencing the divine directly, contact with angels is one of them. Barrès included Poussin but also another French painter, Eugène Delacroix, as such an initiate: "For twenty or thirty years, I have seldom missed a month in which I have not visited Saint Sulpice, in the Chapel of Angels, the famous fresco by Eugène Delacroix, Jacob's struggle with the angel."

For Maurice Barrès, Delacroix was, like him, aware of "our links with those great mysterious beings that link heaven and earth". It was on October 2, 1849 – the Feast of the Guardian Angels in the Catholic tradition – that Delacroix began this series of three paintings on the theme of angels: *Jacob's Struggle with the Angel, Heliodorus Chased from the Temple by the Angels* and for the ceiling's illustration, *The Archangel Michael Who Strikes Down Lucifer.*

It would take him twelve years to complete the work, but it explained – in Barrès' opinion – the key concepts of how to interact with angels:

> [T]he greatest victory is indeed to conquer the angel, to wrestle from him his secret. The angel wants to open for us the gateway to the invisible, it is his mission, but he will not open it without a fight; he does not open it for those who are indolent, tepid, but only to those who, to clear the passage, do not fear to do battle with him.

In short, to have experiences of the divine, one needed to be fearless – like Jacob...

Overcoming the Angel

Jacob's battle is a spiritual one: he addressed his prayers towards an angel who was not inclined to forgive the wrongdoings that Jacob had done towards his brother. The battle illustrates the persistence with which Jacob asks, and asks again and again, forgiveness for his errors.

His prayers lasted an entire night, throughout which the angel could not move on, for he was 'caught' by the prayers that were addressed to him – revealing the first lesson: that an angel has to listen to a human being who is praying.

Late at night, Jacob, rather than waiver, intensified his requests, at which point the angel felt he needed to use his supernatural powers. He touched the thigh of Jacob, expecting him to cringe in pain, unable to continue with "his fight". But Jacob continued to pray.

When the sun was about to rise, the angel had no other alternative but to propose to Jacob that the battle end. But against all expectations, Jacob underlined his determination to receive an angelic pardon for his sins, rather than end the fight. The angel then asked his name. Once informed, the angel replied: "your name will no longer be Jacob, but Israel, for you have the power of a prince with God and amongst men, and you have conquered." At that moment, Jacob demanded to know the name of the angel, who told him it was Peniel. Learning the name of your angel, was another important part of being an 'Angelic Initiate'.

Jacob's story suggests that there is far more to angelic encounters than commonly told. Traditionally, angels are merely seen as messengers conveying the message from God to Mankind, but certain saints – such as Augustine and Gregory – underlined that an angel is above all a function in the sacred hierarchy: they are close to the Throne of God in the celestial court. The role of servant of the divine is expressed in the word "helper", as well as a statement in the Gospel of Matthew that God can always rely on their counsel – on several occasions we are told that there are seven angels with the specific function of being near the Throne of God: the archangels.

Furthermore, in the Old Testament there are also references to angels descending to Earth and having sexual encounters with human beings, the result of which were abominations – which was apparently one of the principal reasons why a Deluge swept across the Earth.

Jacob's Struggle with the Angel, by Eugene Delacroix

Soldiers of the Nine Sisters

Barrès' writings are one of just a few precious resources available to illuminate the enigma of the "Angelic Society". There are two other French writers, Jean Cocteau and Anatole France, who have revealed some key information about it. Jean Cocteau is – maybe not coincidentally – another alleged Grand Master of the Priory of Sion. Cocteau went through an internal awakening to a reality beyond the confines of everyday life in the early years of the 20th century:

> [T]he first sound of the bell, which will finish only with my death, was given to me by Diaghilev, one night, on the place de la Concorde [...] As I questioned him on his reserve (I was accustomed to the praises), he stopped, adjusted his monocle and said to me: 'Astonish me.' [...] This sentence saved me from a brilliant career. I guessed that one does not astonish Diaghilev. From this minute onwards, I decided to die and live again. The work was long and atrocious."

Then, in 1925, Cocteau met his own guardian angel in – of all places – an elevator. He felt a sudden presence, right beside him, of "something both terrible and eternal". This "thing" identified itself: "My name can be found on the plaque." There was only one plaque, and it listed the maker of the elevator: "Heurtebise." The unknown, which for years had been sending its "parliamentarians" to Cocteau, had finally decided to reveal itself. From then onwards, Heurtebise accompanied Cocteau in all of his works. Or, rather, he showed him which road to take and thus guaranteed that Cocteau would follow the path that had been set out for him by the angels – his divine mission. He was now an instrument of the angels here on earth, to fulfil a heavenly plan:

Angel, soldier of the nine sisters
You know what is on the chart
My mysterious way
And as soon as I deviate
You seize me by the hand.

Another 'suspect' for membership in the Angelic Society, and someone who documented its lore, is Anatole France – the author of *The Revolt of the Angels* (1914). Interestingly, France was a friend of Barrès. Anatole France's novel was somewhat autobiographical: he was the son of a bookseller and after working for his father, secured the position of a cataloguer at Bacheline-Deflorenne and Lemerre. In 1876, he was appointed a librarian for the French Senate. France received the Nobel Prize for Literature in 1921, some argue largely thanks to *The Revolt of the Angels*. But apart from a Nobel Prize, in the 1920s France's writings were placed on the List of Prohibited Books, censored by the Catholic Church for being a danger to itself and the faith of its members.

The Revolt of the Angels opens "beneath the shadow of St. Sulpice", in the ancient mansion of the d'Esparvieu family, which contains an elaborate family library. Its librarian and cataloguer begins to notice that each night, a number of books are mysteriously displaced from his carefully ordered shelves, which is beginning to drive the man insane, specifically as it is clear that no human agent can be responsible. The librarian is also friends with Guinardon, a restorer of paintings, whose "favourite subject was the Chapelle des Anges in Sulpice", in which Delacroix's paintings by that time were peeling off the walls, Guinardon being tasked to restore them. Guinardon states that "Michael is my patron saint. And I have a special devotion for the Angels."

It is however the character of Maurice d'Esparvieu who begins to take on a role of primary importance. A man destined to very little in life, d'Esparvieu's fortunes improve when his guardian angel appears. First, the angel tells him that he is the one that is responsible for the disturbance of the library, as he has been studying. He goes on to announce that he has now materialised as he has chosen Paris to prepare the revolt of the angels – not against God, however, but a usurper demiurge, Ialdabaoth. As such, he can no longer be Maurice's guardian angel, for he has a greater mission to accomplish, one that affects all of humanity, if not the universe as a whole.

Intriguingly, the guardian angel is named Arcade – a clear reference to Arcadia, the theme so beloved by the Angelic Society. Arcade stipulates this is his name amongst men: for the angels, his name is Abdiel. Abdiel in Hebrew means "Servant of God" and is identified in the Bible (1 Chronicles 5:15). He was a Seraph in the *Sepher Raziel* and features prominently in John Milton's *Paradise Lost* (1667) – the most likely inspiration for France's tale . However, in *Paradise Lost,* Abdiel denounces Satan after hearing him incite revolt among the angels, and abandons Lucifer to bring the news of his defection to God. His devotion to Lucifer in *The Revolt of the Angels* is never in doubt: his enemy is the Demiurge…and his army of faithful angels, led by Michael.

It also becomes clear that Arcade is not the only materialised angel in Paris and, over the next few weeks, several convene, plan and plot to overthrow the Demiurge from his throne and place Lucifer on it. Three long chapters of the book in fact are a 'history lesson' in the history of the angels, which includes a detailed account of their first attempt, before time began, to overthrow Ialdabaoth from his throne – which had caused Lucifer's fall.

With the preparations all in place, Arcade and his fellow angelic conspirators go to Lucifer's place of exile on Earth, to inform him that his army is ready to fight for the throne. However, despite having a dream in which his army is able to conquer the Demiurge and expel him from heaven, Lucifer wakes up stating he will not fight. Instead, he believes that we need to fight and conquer Ialdabaoth within ourselves!

Return to the Golden Age

France's book was a novel, but the scenario it described was something members of the Angelic Society apparently believed: that in the past, Lucifer had perverted the angelic realm, which had disastrous results for humanity. That in itself is nothing new and straight from Christianity. But what differentiated members of the Angelic Society from the rest of humanity was that they strived to right this wrong, and restore the world to the state it had been in before, with the help and guidance of the angels themselves. The mission of the Angelic Society was the return to a Golden Age – also known as Eden or Arcadia; a "Time Before Time".

To understand what this means, we need to turn to 16th and 17th century England. There, we find the likes of Edmund Spenser and especially his friend, Philip Sidney, writing on the theme of Arcadia, underlining that – unsurprisingly – contact with the angelic realm was not limited to French soil. The most notorious 'contactee' was probably John Dee, whose antics require a far deeper treatment than is possible here. However, the answers we seek can be found Milton's *Paradise Lost*, which unveils what this 'original state' was.

In Book IX, Milton explains that the Fall of Mankind had geological repercussions for the Earth – the angle of the Earth's axis went from being upright to the present 23.5 degrees:

> Some say, he bid his Angels turn ascanse The poles of earth, twice ten degrees and more, From the sun's axle; they with labour pushed Oblique the centrick globe...

'Askance' means 'sideways' or 'deviate,' and we are told that "he" – meaning God – had "bid his Angels" to do this. That is, he ordered that they tilt the earth and push it away from "the sun's axle" – the upright ecliptic pole. The rest of Milton's verse explains why:

> [T]to bring in change of seasons to each clime; else had the spring perpetual smiled on earth with vernant flowers, equal in days and nights, except to those beyond the polar circles; to them day Had unbenighted shone...

Today, the equinoxes are the only two times of the year when the Earth's axis is upright and the Earth is in the image of the "Time Before Time". Perhaps unsurprisingly, the equinoxes therefore became extremely important in the works of some artists, such as in Botticelli's *Primavera*. Interestingly, the original title of this painting was *The Time Returns*, providing a further insight that Botticelli may have at least known of the Angelic Society's tradition. Botticelli was part of the 15th century Italian Renaissance, which saw a revival of Greek thinking – "the Greek ways" – during which angels once again became prominently depicted in works of art. Most importantly, amongst Botticelli's friends was Michelangelo, who was named by his parents after the Archangel Michael – an extremely rare occurrence – and who would become one of the greatest artists ever. One can only wonder whether his hands were moved in a similar angelic vein as Andrew Lakey. The sheer mastery of his work suggests that they were.

The scope of doing research into the Angelic Society is therefore vast. Even for France, we have only touched upon some 19th and 20th century artists. But it is clear that the Angelic Society consisted of some of the most famous artists the world has ever seen. And they likely excelled in their arts because of their angelic connection: their art truly was from another realm. But the Angelic Society was never an organised secret society, like the Freemasons, Rosicrucians, or like. There are no membership lists. They were a society in the sense that all of its members share(d) a common bond – their lives were guided by angels – and they fulfil(ed) a divine mission here on Earth. Together with the denizens of this other realm, they worked for the return of a Golden Age, a time before the Fall; they were here to right one of the most famous wrongs of all times – Lucifer's Rebellion –

yet their endeavours are hardly known at all...their art is the only testimony to their angelic mission.

Philip Coppens is an author and investigative journalist on topics ranging from the world of politics to ancient history and mystery. He is the former editor-in-chief of the Dutch magazine *Frontier* and is a frequent contributor to various publications. Since 1995, he has lectured extensively across the world. He is the author of *The Stone Puzzle of Rosslyn Chapel, The Canopus Revelation, Land of the Gods, The New Pyramid Age,* and the recently released *Servants of the Grail.* His website is www.philipcoppens.com.

"The Invincible 45" (Alex Andreev, www.alexandreev.com)

Return of the Flying Saucers

a fresh look at the sighting that started it all

by *Martin Shough*

The sighting by pilot and businessman Kenneth Arnold of nine "peculiar aircraft" skimming over the peaks of the Cascade Mountains of Washington in June 1947 has unique significance. The worldwide press coverage that followed defined for all time the public consciousness of "flying saucers", and we will never understand their historical origins without an understanding of what Arnold said he saw and of how his story was interpreted in the context of the times.

One aspect of this question concerns Arnold's fitness to see what he said he saw and to accurately describe it to us in the days and weeks afterwards. Was Arnold a "reliable witness"? What does that mean? What is this quality of reliability? Is there any way of measuring it?

Another aspect is how the event and the reporting of it changed Arnold himself, along with the world around him. Arnold's own descriptions of the objects he saw in June 1947 altered over the years. The history of this change is complicated and the reasons for it difficult to extract. Why should this confusion have arisen? How ought we to deal with it?

There is a cynical and rather lazy point of view which holds that where inconsistent variations arise in witness narratives this merely serves to prove that testimony is always worthless and can tell us nothing. The history of science clearly shows that this is not so, but it is also clear that human observers are socially-embedded and highly sensitive instruments whose fluctuating outputs need to be calibrated with cunning. This is especially true where the embedding psychosocial medium is as richly evolving and as highly cathected as is the flying saucer mythology. For this reason, where the historical record shows that a witness's statements mutate into explicitly inconsistent forms over time we do normally require a strong justification for giving greater – or even equal – weight to the later forms. Logic and experience tell us that we will normally minimise corruption and contamination if we begin with a presumption in favour of contemporaneous evidence.

Arnold's case is the prime exemplar of this general rule. It is unique inasmuch as his initial sighting report was born naked, as it were. By definition no saucer mythology yet existed. But as time went by, what Arnold saw and said he saw became entangled with what society at large came to believe that Arnold saw. As the infant story was handed round to be inspected by admirers and detractors, it became swaddled in mythic embroideries and confusions, until in the end Arnold himself apparently disowned his own offspring and rewrote his will in favour of an imposter. We need to understand how and why this happened.

Bamboozled by Boomerangs…

It has become a widely retailed legend that Arnold never described disc-like objects at all. Many modern accounts assert that he originally reported nine "boomerangs" or "crescents", but that a description of their motion – "like saucers skipped over water" – was misinterpreted by a journalist who thus invented the totally fictitious image of "flying saucers". The journalist responsible has widely been identified, even in some quite recent literature, as Bill Bequette, author of the original story that went out on the AP wire from the Portland *East Oregonian* on June 25 1947.

The true part of this legend is that Arnold did indeed claim, years later, that he had offered the simile of saucers skipping over water as a description of the objects' motion. But the rest is a can of worms.

Although several different motion similes appear in early published sources, and in Arnold's own Air Force report, it should be noted that the "skipping saucers" image is nowhere among them. The original sources contain other motion similes: "like the tail of a Chinese kite, kind of weaving and going at a terrific speed"; "they flipped and flashed along"; "they flew like many times I have observed geese to fly in a rather diagonal chain-like line as though linked together"; "like fish flipping in the sun"; and "like speedboats on rough water".[1] The claim that they flew "like they take a saucer and throw it across the water" doesn't appear in the record until Arnold offered it 3 years after the sighting in a 'phone interview with radio broadcaster Ed Murrow in 1950:

> …when I described how they flew, I said that they flew like they take a saucer and throw it across the water. Most of the newspapers misunderstood and misquoted that too. They said that I said that they were saucer-like; I said that they flew in a saucer-like fashion.[2]

Three days later a United Press wire story out of Boise, Idaho, quoted an interview with Arnold that day (April 10 1950) in which Arnold again complained, that…

> …"the press misquoted me when they said I described the objects as flying saucers." Arnold said he merely described the objects in flight as appearing to skim through the air like a saucer over water.[3]

If "most" newspapers had misquoted him in 1947 then there should be at least one that didn't. But apparently *all* of the papers misquoted him. From day-one Arnold's story was sought by phone and in person by countless reporters who "came out of the woodwork",[4] so one must assume that he had opportunities to supply clarification. Yet the "misunderstanding" was widespread in the media within a few days and stayed there.

The early press certainly reported Arnold's frustration with how his story was being mistreated. One interviewer said that a "harassed" Arnold "sighed" about all the "hoopla and hysterics", complaining: "I haven't had a moment of peace since I first told the story… This whole thing has gotten out of hand. I want to talk to the FBI or someone. Half the people I see look at me as a combination Einstein, Flash Gordon, and screwball. I wonder what my wife back in Idaho thinks."[5] Nevertheless not a single early source reports Arnold protesting that his use of the word "saucer" had been misunderstood, or reports the statement that they flew like saucers skipped across water; whereas several sources, including Arnold's own AAF report, *do* contain statements – contrary to Arnold's late claim – that they were saucer-like in shape.

The original press stories were written by Pendleton *East Oregonian* journalists Bill Bequette and Nolan Skiff. The phrase "flying saucers" appears in none of them but was invented by an unknown journalist or editor elsewhere (probably about June 27) on the basis of Bequette's wire stories.

Bequette and Skiff had a first interview with Arnold in the newspaper office about noon on the morning of June 25, after which the initial stories were quickly written. The very first brief story by columnist Nolan Skiff, written just in time to make the bottom of the front-page of that day's issue of the *East Oregonian*, uses the phrase "saucer-like aircraft", proving that right from the start Skiff interpreted Arnold's use of the word "saucer" that morning to be a shape simile.

Bequette had suggested to Arnold that a wire story might shake loose some information about the strange objects which both he and Arnold assumed were some sort of Army Air Force planes or rockets. He wrote a separate, slightly longer story which he put out on the Associated Press wire at the same time. Consistently with Skiff's story it, too, said that Arnold (mistakenly identified as a US Forest Service employee) had described seeing "nine bright saucer-like objects".

At this point the two journalists went innocently to lunch. When they got back they were surprised to find the office secretary struggling to field telephone calls and messages from all over the country demanding more information. Bequette's AP wire seemed to have stirred up the entire Fourth Estate and he realised that he had misjudged the story's impact, so "I had to hustle down to the hotel, find Arnold, and wring out every last detail."[6]

Bequette spent a further two hours interviewing Arnold at his Pendleton hotel that afternoon. A follow-up article appeared in the *East Oregonian* the next day, June 26, and was also phoned through to Portland whence it went out on the United Press wire. It naturally contained much more detail, and corrected the error about Arnold working for the Forestry Service; but conspicuously it did not correct the use of the word "saucer" given as a shape simile in Skiff's article and in Bequette's own earlier AP wire story the day before. The new story not only repeats the simile, but this time puts it explicitly into the mouth of Arnold himself, who is now quoted describing the objects as "saucer-like".[7]

Bequette also quotes two different motion similes offered by Arnold. Neither of these is the "skipping on water" simile that Arnold much later claimed to have given to Bequette. Instead Bequette first quotes Arnold as saying that their erratic motion was "weaving like the tail of a Chinese kite" and then adds, "He also described the objects as 'saucer-like' and their motion 'like a fish flipping in the sun.'". We should note that Bequette here *explicitly separates* the shape descriptor 'saucer-like' from an associated motion simile (one which Arnold also used elsewhere).

When questioned by sociologist Pierre Lagrange in 1988 Bequette evidently did not remember the "skipping saucer" motion simile; neither did he believe that he had coined the phrase "saucer-like" as a shape simile himself. His original story had placed this phrase in quotes and attributed it to Arnold. But he told Lagrange that it was possible and that he was prepared to give Arnold the benefit of the doubt as to what he had meant.[8] However when speaking to author Ronald Story in early 1992 his memory seemed clearer on this point, saying that Arnold *had* used "saucer" as a shape-simile that day. Cognizant that Lagrange had recorded a less explicit answer, Story remarked: "I can only repeat what he confirmed to me: that [it] was based on Arnold's description."[9]

The record tends to support Bequette's memory. In addition to the Bequette and Skiff stories there are several other early news sources quoting Arnold in the same terms, including for example further news service wire reports out of Pendleton by unnamed reporters on the 25 and 26 June. One UP dispatch quotes a local businessman to whom Arnold had described the objects as "shaped like saucers".[10]

A "special" correspondent for the Chicago *Tribune* filed a story after an interview with Arnold on June 25, quoting Arnold as saying that the objects were "silvery and shiny and seemed to be shaped like a pie plate". There is no mention of plates "skipped over water". The tableware simile appears only for the shape, not motion. As regards motion Arnold is quoted as saying that they "weaved in flight like the tail of a kite" and "went by me like a bullet."[11]

Of course press stories can be incomplete. The only early source where we can be absolutely certain that we have all of Arnold's own words accurately recorded without loss of context is the KWRC radio interview of June 26 1947. Two things about this broadcast are notable for our present purpose. The first is the background of press activity well-described by host Ted Smith, who indicates that Arnold had been interviewed directly for United Press wire reports by UP staff in Portland and perhaps elsewhere, as was Smith himself:

> Well, Kenneth, thank you very much. I know that you've certainly been busy these last 24 hours, 'cause I've spent some of the time with you myself, and I know that the press associations, both Associated Press and our press, the United Press, has been right after you every minute. The Associated and the United Press, all over the nation, have been after this story. It's been on every newscast, over the air, and in every newspaper I know of. The uh, United Press in Portland has made several telephone calls here at Pendleton to me, and to you this morning, and from New York I understand, they are after this story...[12]

Arnold also tells us himself in the interview that he had already given his own story directly to Associated Press, not just second-hand *via* the wire and telephone reports that we know were sent by Bill Bequette on 25 and 26 June. This reinforces our impression of the sort of opportunities that were available for Arnold to correct a press misapprehension directly to the wire services, had it really been the case that he had reported nine "boomerangs". And the second point: In this definitive early source, broadcast on June 26, Arnold does not mention anything about "skipping on water", neither does he correct Nolan Skiff's *East Oregonian* article of the day before attributing to him the description "saucer-like aircraft". He was not obliged to do so, of course; but it was another opportunity and we can be certain that he did not take it.

We also find that Arnold himself used both "saucer like objects" and "saucer-like discs" as shape-similes in his own original Air Force report typed by his own hand on or about July 08 1947. Once again, just as important as the fact that Arnold uses these phrases is the conspicuous fact that he does *not* use these terms in the context of any motion simile. Even if previous attempts to correct journalists' misapprehensions had failed – indeed *especially* if they had failed – here was the opportunity, two weeks on, for Arnold to set the record straight first-hand in the most important, official forum. But far from taking the chance to explain that he only mentioned saucers in the first place in order to suggest a skipping motion, Arnold explicitly confirms "saucer" and "disc" as shape similes, whilst in reference to motion he says only that they "flew like many times I have observed geese to fly in a rather diagonal chain like line", and erratically "dipped" or "flipped and flashed in the sun" as they "swerved in and out of the high mountain peaks."[13] And even though the shape of the "disk" Arnold drew in this report (and repeated numerous times on carbon copies for other people) was more shovel-shaped or shell-shaped than truly saucer-shaped, with an axial ratio "longer [in the direction of motion] than wide", it most emphatically was not remotely crescent- or boomerang-shaped (which would of course have been wider than long).

...And Bewildered by Bats

There are two oft-cited early references that appear to be significant exceptions to the dominant discoidal description. One early newspaper report does use the phrase "crescent-shaped"; another quotes Arnold as describing the objects as "somewhat bat-shaped". We will consider these in turn.

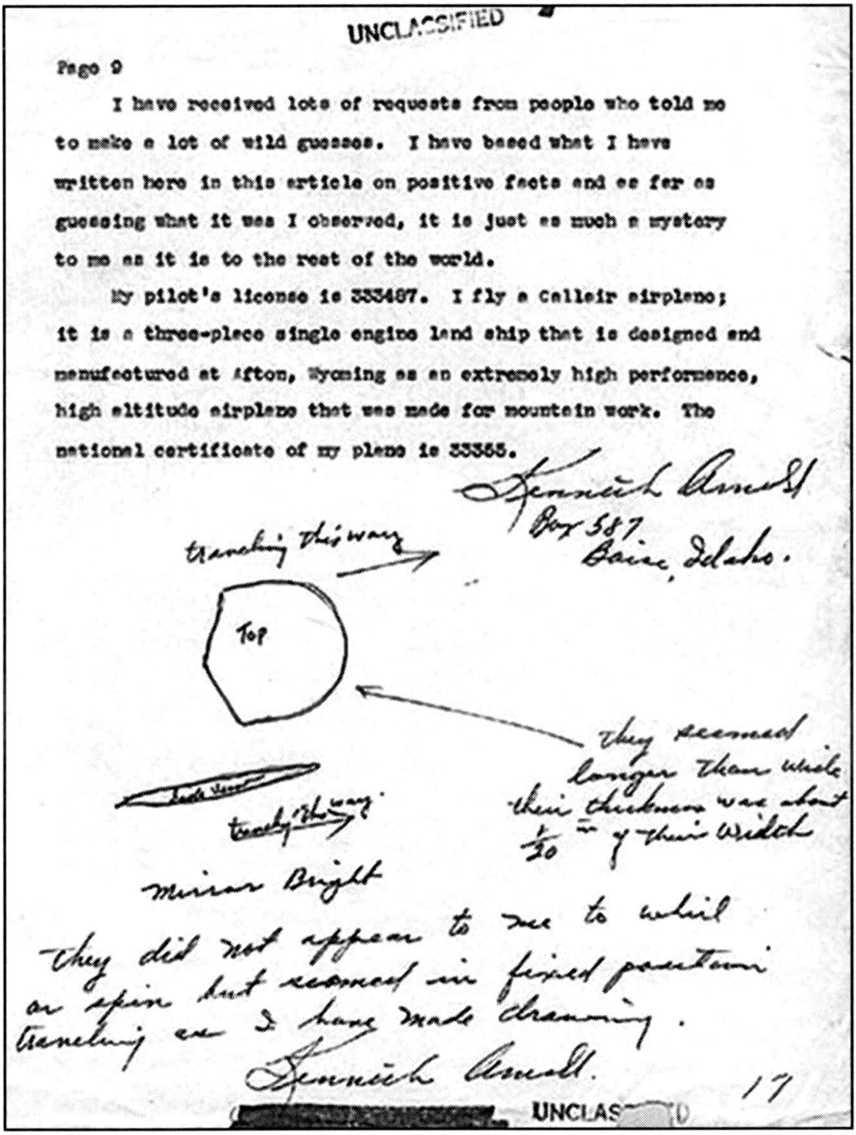

UNCLASSIFIED

Page 9

I have received lots of requests from people who told me to make a lot of wild guesses. I have based what I have written here in this article on positive facts and as far as guessing what it was I observed, it is just as much a mystery to me as it is to the rest of the world.

My pilot's license is 333487. I fly a Callair airplane; it is a three-place single engine land ship that is designed and manufactured at Afton, Wyoming as an extremely high performance, high altitude airplane that was made for mountain work. The national certificate of my plane is 33355.

Kenneth Arnold
Box 587
Boise, Idaho.

they did not appear to me to whirl or spin but seemed in fixed position traveling as I have made drawing.

Kenneth Arnold.

17

UNCLASSIFIED

Page from Kenneth Arnold's original report to the Air Force showing 'shovel' shape

The *Oregon Journal*, June 27, said that Arnold "clung stoutly to his story that he saw nine shiny crescent-shaped planes", but these words are not in quotes from Arnold, they are the writer's. Where Arnold is actually quoted in the same article he says, "They were *half*-moon shaped, oval in front and *convex* on the rear. I was in a beautiful position to watch them…they looked like *a big flat disk* [emphases added]." This describes the sort of shape Arnold drew for the Army Air Force, a flat plate with a trimmed off or tapered rear edge, and the "half-moon" clearly plays the same role here as the "half pie-pan" in the description used by Arnold elsewhere: "half a pie-pan with a convex triangle in the rear". The shape in Arnold's drawing suggests that he may have had in mind a *gibbous* moon, i.e. between half and full; howsoever the reporter has interpreted "half" to mean "crescent" (in some people's imaginations "moon" and "crescent" might be almost synonymous) and neglected the rest of the description.

The other phrase is to be found in Bill Bequette's second story published in the *East Oregonian* and telephoned to Portland on June 26 (see above). In this case the phrase appears in quotes, and at first sight is more troublesome. Arnold described the objects to Bequette as being "flat like a pie-pan and somewhat bat-shaped". A modern reader whose mind-set is influenced by the crescentic flying-wing imagery which progressively took over during Arnold's later years (beginning with his August 1947 claim that just one of the objects had been been a sharp-winged crescent, and ending with the late claim that all nine had been crescent-shaped) tends to interpret "somewhat bat-shaped" as indicating flying mammals of the genus *chiroptera* which would seem to imply wide, extended wings. But one's first impression is frustrated by the fact that Arnold is also quoted in the same interview as saying that the objects were "saucer-like".

The apparent incongruity of a *chiroptera*-like image among the other descriptions and drawings of June/July 1947 may be a matter of our excessive literalism. Researcher Barry Greenwood has pointed out[14] that the *Batman* comic-book franchise was popular at the time and that "bat"-inspired design was widely disseminated in print and in other merchandise. This imagery was often far from naturalistic. Such imagery might have influenced Arnold to think of his saucers' pointed trailing edges as "somewhat bat-shaped".

Another interpretation of that early phrase might be that Bequette quotes the phrase out of context. Other possible explanations then are that Arnold was referring to table tennis bats and/or aircraft-marshalling signal bats.

Table tennis bats are widely (although not universally) known in the US as "paddles". A table tennis bat, not quite round and with its "tail" edge tapering towards the handle, could well be an apt simile for the objects Arnold described and drew, but an American in 1947 would arguably have been much more likely to think of the word "paddle" than the word "bat" in this context.

Perhaps for Arnold, as an aviator, a more likely allusion would be to the signal bat used in aircraft marshalling during that era. This object, "a flat round club with a short handle, resembling a table-tennis bat, used by a man on the ground to guide the pilot of an aircraft when taxiing"[15] appears universally to have been known as a "bat", and the activity was known as "batting". Arnold may well have been familiar with this practice of "batting" from the ex-Army Air Force flyers that he mixed with as well as from newsreels and general aviation-community scuttlebutt.

Looking at the famous crescentic flying wing illustrated in his 1950 pamphlet and the 1952 book (an artist's rendering of a wooden model Arnold said he made for the AAF) many have assumed that this explains an early use of the phrase "somewhat bat-shaped", but this is a misunderstanding based on a failure to carefully examine date order. The flying wing model represents only *one* of the objects, and Arnold's own first realisation of its unique shape postdates by at least five weeks his use of that phrase, which he used only in the context of his earliest "disc" and "saucer" descriptions.

Arnold showing crescent-shaped UFO

By his own account Arnold did not tell a soul that a ninth object might have differed from his "saucer-like objects" and "big flat discs" (i.e. might possibly have been wing-like, crescentic) until saying this privately to the two Army Counter-Intelligence Corps officers Capt. Brown and Lt. Davidson who interviewed him on July 31 1947.[16] Reflecting on this omission when first discussing the matter publicly in 1952 he makes a point of excusing it, explaining that it had been an impression too uncertain even to mention to his wife. This is all clearly inconsistent with the theory that he had told the newspapers about a *chiroptera*-shaped crescent wing as early as June 25 1947.

Moreover Bequette's June 26 quote implies that *all* the objects ("they") were "somewhat bat-shaped". The flying crescent model reflects accurately the outline he had drawn for the FBI in Aug 1947 of just one of the objects, "a very wraith-like looking thing [that] wasn't round at all",[17] which he said he selected for the purpose because it had seemed distinctive. That early drawing was made a couple of weeks after Arnold had been exposed to Capt Brown's drawing of the Rhodes object, and to Lt Brown's description of it as a "flying wing", and it was captioned in his own handwriting as follows: "one object looked like this, the second from last of the formation. Seemed slightly smaller." Plainly the shape of this newly-remembered odd-man-out has nothing to do with the phrase "somewhat bat-shaped" used to characterise all of the objects weeks earlier when Arnold had not even been clear in his own mind that there had been an odd-man-out at all. Equally plainly Arnold must have intended the phrase to convey something about the "big flat discs" with their tapered trailing edges.

Once the story of the single flying wing emerged Arnold appears to have maintained it consistently in public for many years. As far as can be determined he never described this one object as bat-shaped. The allusion to bats vanishes early from the record, before Arnold even acknowledged the existence of the flying wing. It was apparently a simile he used once that day in 1947 and, unlike "saucer" and "disc", never repeated. In later years he does not speak of mammalian bats, but rather of manta rays, rafts, mediaeval axes and other things.

In the end we cannot know for sure what Arnold meant when he said whatever it was he said to Bequette and to other journalists and intermediaries in June 1947, only what was printed, or broadcast and recorded. But, however one looks at it, it is hard to avoid the conclusion that the term "saucer" must have been chosen by Arnold that day to imply something significant about the shape of objects that he also described as featureless mirror-bright "big flat discs", objects which, whilst not truly circular, were somewhat round.[18]

And although from 1950 Arnold denied having used the word "saucer" exclusively or exactly as a shape simile, this is not the same as a denial that the objects were discoidal. A preponderance of evidence shows that he *was* prompted to use the word "saucer" *at least in part* as a shape simile for his "big flat discs".

Arnold apparently never believed or claimed that his objects were *perfectly circular* discs, but his reassertion of this fact is too often presented as though it were equivalent to an admission that he had really said that the objects were boomerang shaped. I have found no source where Arnold ever described any of the objects as "boomerangs" and the uncritical re-echoing of this claim throughout much recent sceptical literature has been damaging. Nevertheless, the combination of vagueness and image-creep in Arnold's own descriptions post-1950, culminating in his denial of having ever employed a "saucer" shape simile at all, are at least partially responsible.

Of Flying Discs and Flying Wings

Would Arnold have had any special reason to expect to see discoidal aircraft in June 1947?

At that time images of approximately-discoidal aeroforms were not totally unknown. It is well-known that a few of the hundreds of imaginative spaceships of all shapes depicted in cover art for *Amazing* and similar publications during previous decades did have discoidal symmetry. But Arnold, according to his own account, had no familiarity at all with sci-fi and fantasy magazines, and certainly did not interpret his sighting in a fantastic context. He thought he had seen Army "aircraft" or perhaps "guided missiles".

The idea of disc-like aircraft did exist on the fringe of the aeronautical world. A speculative item about a circular-wing aircraft design had appeared in print in the pulp magazine *Amazing Stories* in 1946. The magazines *Science et Vie* in France and *Mechanix Illustrated* in the US had both carried cover pictures of the USAF's abortive experimental plane – the heel-shaped "Flying Flapjack", only one of which briefly flew in Connecticut before being scrapped - during the previous year. But there is no evidence in any document that Arnold had heard of these ideas at the time, still less that he was influenced by an interest in them.

In contrast to this, flying wing designs were more than a fringe idea and a fond hope - they were an engineering reality, developed in

Vought V-173 'Flying Pancake', predecessor to the 'Flying Flapjack'

several forms in the US by ex-Lockheed designer Jack Northrop since 1939 (from a concept pioneered by Northrop in 1929) and famously in Nazi Germany by (among others) the Horten brothers, whose designs had been the subject of much interest by Britain and the US in the immediate post-war period. This image was far more widely disseminated in the popular culture.

Indeed one witness (a Forestry Service look-out in the Cascades who saw a line of bright somethings on the same day as Arnold) thought it prudent to emphasise to journalists that what he saw "was not the flying wing";[19] whereas Clyde Homan, manager of a tulip-growing business who along with his farm foreman saw nine similar objects undulating and giving off sun reflections as from bright metal near Woodland, Washington on June 27, "ventured the opinion the objects might have been the new type of tail-less aircraft known as flying wings", even though he could not make out any shape behind the bright reflections except that they were "very flat and very, very thin".[20] A couple of days later a Portland newspaper quoted the assurance of Col. Carl Spaatz: "The Army has no aircraft that could fit description of the discs; it is not the flying wing."[21] A

3-page illustrated spread on the Northrop XB-35 had appeared in the January 1947 issue of the widely-read magazine *Popular Science.* It would probably be fair to say that the flying wing was the iconic image of futuristic aviation in post-war America.

If Arnold's sighting had been influenced by notions of what advanced planes ought to look like then one would expect his report to have emphasised features of flying wings rather than of flying discs. And there *are* features of Arnold's early verbal description that, lifted out of context, can be interpreted as suggestive of flying wings. He spoke of the objects' "wing or whatever it was" and emphasised that what especially puzzled him about their shapes was that he "couldn't find any tails on them". If we ignore talk of saucers and discs, and if we ignore also Arnold's own drawings, with their annotation saying that the flat discs were "longer than wide", then his description of "half a pie-plate with a convex triangle in the rear" could also be interpreted as a kind of flying wing (wider than long) with a small "triangle" being a sort of vestigial fuselage, and from this we could get to Arnold's later image of the sharp-winged crescent. Is it possible that these details are, as it were, fossils of Arnold's true first impression, preserved inside a more discoidal image which Arnold improvised during the first weeks as he subtly adapted his story to conform to popular expectation?

It is understandable that there may have been a cultural pressure in this direction, and one can point to individual influences. Arnold's discoidal sketch in his Air Force letter must have been drawn after the July 04 United Airlines case which is referenced in the letter. By July 04 the "flying saucer" or "flying disc" stereotype is becoming well established everywhere, and Capt. E. J. Smith – an impressive and influential witness with whom Arnold closely allied himself in his appeals to the Army Air Force to take the saucers seriously – had that day seen nine "discs" described as "circular, flat on the bottom and rough on top, bigger than our [DC-3] aircraft" silhouetted against the Idaho sunset.

So here we have a theory: Perhaps in Arnold's post-July 04 sketch an original, more idiosyncratic impression of a tail-less flying wing was morphing towards a "saucer-like disc" because this stereotype was taking over the public imagination and had even influenced Capt. Smith, regarded by Arnold as "probably the most highly thought of and respected veteran pilot that flies the air lanes".[22] Perhaps Arnold was motivated by wanting to hang onto the coat tails of "Big Smithy" and borrow the support of the growing number of other "saucer" eyewitnesses around the country?

But this suggestion seems out of character for Arnold, who was no meek hanger-on but a tall, well-built and capable self-made man with a healthy opinion of his own worth. He gives the impression of sticking stolidly to his guns in the midst of speculation and ridicule. Character testimonials from Army Air Force Counter-Intelligence Corps agents, journalists, aviation professionals and others who knew and/or interviewed him paint a picture of a self-reliant and forthright man not inclined to tell people what they wanted to hear. The theory that in the first couple of weeks he transformed flying wings to flying discs to ally himself more closely with a popular "flying saucer" craze is not very consistent with the character and actions of a man resentful of being characterised as a "screwball" and who complained, "I haven't had a moment of peace since I first told the story… This whole thing has gotten out of hand. I want to talk to the FBI or someone."[23]

This invites us to consider an alternative scenario in which Arnold perhaps tended to revise an original quasi-discoidal description in the other direction, away from an image increasingly associated with wild speculation and towards more aeronautically-credible flying wings, by increasingly emphasising features which had distinguished their shapes from true discs and/or suppressing explicit statements that might have suggested circular symmetry.

Arnold certainly had motivation. His initial efforts to get the AAF and FBI to take him seriously seemed to be frustrated, and he expressed disappointment about this in his original AAF letter and in a telex of July 12 1947 to the Public Information Officer, Wright Field. Official reticence may have increased his aversion to being publicly associated with saucer-sighters whom the newspapers too often portrayed as nuts and kooks.

On July 31 1947 Army CIC officers Brown and Davidson exposed Arnold to what they called a "flying wing" photographed by William Rhodes in Phoenix, intimating that it was "genuine". Arnold's reaction is consistent with a tendency to seek the endorsement of conservative military authorities. He suddenly remembered a fact that he had never mentioned before – that one of his objects had looked just like it. It would also fit the way the explicit descriptions and drawings that are part of the public record prior to this date give way later to vaguer and more evasive descriptions. His 1950 pamphlet "The Flying Saucer As I Saw It" and his 1952 book are both notable for a reluctance to be verbally explicit about shape, and both allow the image of a sharp-tipped crescent wing to stand alone as representing *at least* one and – but only by tacit implication at this stage – possibly all of the objects.

Arnold told the Army Air Force that he had spoken with former wartime AAF fliers who had been alerted about radical jet designs that might be encountered in the European theatre – an apparent reference to experimental Nazi aircraft. And Arnold was quite proactive in pursuing the mystery of his sighting (*vidé* his activities in the Maury Island affair and contacts with the AAF), which raises the possibility that Arnold might have been made aware soon after June 24 of rumours of possible recovered Nazi designs. Could he have learned or guessed that these designs included flying wings, of interest not only to US engineers hoping to build on the initial promise of the Northrop XB-35 flying wing but also to the Russians?

The idea would feed into his belief that he saw secret AAF planes or rockets. He could well have come to believe that this was the only reasonable explanation for the AAF's reluctance to investigate. And

Northrop XB-35 'Flying Wing'

a flying wing could well have been the image in his mind when he began "adjusting" his recollection of the the objects' shapes towards something aeronautically plausible that he knew the AAF would be more inclined to take seriously.

By 1977 Arnold had inverted his pre-1952 account of eight "plainly observed" discs and one half-glimpsed "smaller" flying wing, to describe, instead, one "*larger*" flying wing, darker and now *clearer* in outline than eight smaller objects of more uncertain shape, hinting that the rest might have been crescent wings too.[24] By Feb 1978 he was telling journalist Bob Pratt that all of the objects had been crescent-shaped wings.[25]

For a man of Arnold's character – who thought of himself as one of a community of no-nonsense mountain aviators and whose daily milieu consisted of other fliers, some of whom were ex-Army – a desire to court the respect of conservative authority figures in the military is somewhat more psychologically plausible than a desire to reinforce a media reputation as "Mr Flying Saucer", and such a desire could have influenced him to morph the shapes of his objects progressively towards flying wings, *via* the "genuine" Rhodes object, which was drawn for him by Army CIC agent Capt. Davidson and described to him by Lt. Brown as a "flying wing".

The alternative is that Arnold initially believed he saw nine flying wings but subsequently told the Army Air Force, Army Counter Intelligence and the FBI that they were (with one remembered exception) discoids in order to fall in line with the cultural predominance of a "saucer" motif invented mistakenly by journalists. This really does not fit either the psychosociology of the moment or the historical documentary evidence.

In summary, it appears that Arnold really did originally report what could be fairly described as nine distorted saucers, which he himself described as "saucer-shaped", "saucer like" and "big flat discs", not exactly circular like mother's crockery but (as he drew them multiple times) somewhat round shapes that had "looked perfectly circular" when seen by the reflection of the sun off their mirror-like tops, but which were revealed as roughly shovel-shaped or scallop-shaped when "observed quite plainly" in black silhouette against the snow, having a very shallow, dished cross-section when seen edge on. And the most likely explanation of his later drift towards a crescent description is that he regretted using the "saucer" shape simile, and denied it, because of unintended consequences: He wished to distance himself from the risible *cultural stereotype* of the "flying saucer".

Was Arnold a 'Credible Witness'?

Kenneth Arnold is the one witness to "flying saucers" whom we know could not have had any prejudicial prior expectation of seeing them, for they didn't yet exist. This makes Arnold's observation almost uniquely uncontaminated.

But of course no observation is a passive process taking place in a psychological vacuum. Observation is always an active process of filtering and pattern-matching controlled by an individual's mental set. There is no sign of futuristic fantasy or space preoccupations in Arnold's mental set, but it transparently included factors like aviation, the post-war politico-military situation and ornithology, and we can see these operating on his expectations at different points in his accounts of the sighting. These factors can to some extent be allowed for. But how can we know that, in addition to these explicit prior determinants, there may not be others, unidentified but still likely to distort the output of the observing instrument called Kenneth Arnold in unpredictable ways?

One avenue is to cross-check qualitative and quantitative details of the original reports for internal consistency. The reasoning here is that when an event is projected from the physical to the reportorial through, as it were, the lens of psychological preconception, there is the opportunity for distortions which skew the overall picture, and these distortions are not required to be symmetry-preserving. In other words the projected picture is not required to make physical and geometrical-optical sense. On this internal consistency test Arnold scores quite well.[26] But this only tells us that if Arnold's observation was significantly distorted by unknown psychological factors they have left no obvious internal trace.

Another avenue which I have been urged to take by several correspondents[27] is to apply lessons learned from studying Arnold's experiences and actions *after* the event. Arnold had a number of other minor sightings in the months and years after June 24 1947, and it has been argued that this fact reveals him to have been an unreliable observer, prone to error.

In a similar vein, his activities as what has sometimes been called "the first ufologist" in the Maury Island affair are often cited as tending to indicate poor judgement or worse.

These issues are more difficult to reason about because there is so much subjectivity involved. As investigators we, too, each have our own 'mental set' that predisposes us to favour one interpretation or another, just as Arnold did. This is why it is important to test our conjectures as far as possible against specific documented facts, and to take care to respect the historical time order of these facts.

Arnold as 'The First Ufologist' – Maury Island

The Maury Island affair began with a claim that harbour patrolmen had seen doughnut-shaped saucers over Maury Island in Puget Sound, Washington, just a few days before Arnold's sighting. Fragments of material dropped by one of the saucers was said to have been recovered. *Amazing Stories* publisher Ray Palmer wrote to Arnold, now nationally quite famous, and offered to pay his expenses if he would fly up to Tacoma and investigate the story. Arnold went. Should a sensible person have taken up this proposal from a pulp magazine publisher to fund an investigation into the story in the first place?

Arnold ignored the first letter he received from Palmer a few days after his sighting because he had no idea who the man was and didn't read this kind of literature. He was more interested in trying to raise the interest of the AAF and FBI. But he seemed to be failing. He had been asked indirectly for a written report, but had had no feedback and no one had been to see him. After a week a second letter from Palmer arrived in early July, offering payment for a story. This time Arnold responded, but not until the end of the month, and only to the extent of sending Palmer a carbon copy of his letter to the AAF.

Meanwhile, by July 12 Arnold had expressed his "considerable disappointment" to Army officials in writing that they weren't able to offer any explanation of the report he had sent "some days ago". His tone suggests that he was slightly miffed about being given a cool brush off. He then received a third letter from Palmer, this time suggesting he go to Tacoma. Arnold did not reply but put it to one side to "think it over".

Only after this did he at last receive a personal visit from Capt. Brown and Lt Davidson from Hamilton Field, California. "I couldn't figure out," he said, "why such an efficient body as Military Intelligence hadn't called on me before." Now that they had arrived he was impressed by their courtesy, but a meeting with Capt. Smith and *Idaho Statesman* aviation editor Dave Johnson was a disorganised affair from which "nobody found out much". The two Intelligence officers then went through all of the saucer mail received by Arnold since June 24, "much of which I did not feel capable of evaluating", in particular collecting all the requests from organisations for accounts of his sighting. They left him with his curiosity still unsatisfied, but with the invitation to call or wire Hamilton Field if anything else came up.

Soon after this Arnold thought over Palmer's letter and consulted Dave Johnson on the wisdom of taking up the offer to go to Tacoma and investigate the rumoured saucer fragments. Johnson could see no harm in finding out if the offer of funds was sincere and advised him to answer. Arnold did so and, finding himself unexpectedly in prompt receipt of $200, flew to Tacoma on July 29. He did it because he was fascinated and frustrated. He wanted to get to the bottom of what had happened to him. His innocent sighting report seemed to have turned the world crazy so that he seemed "the only sane one of the bunch!" He was unhappy that he was viewed by many of the public as a "screwball" and not taken seriously enough by the authorities. He wanted solid evidence and here was a chance to get it.

Arnold's account of subsequent odd events in Seattle and Tacoma – the pre-booked hotel room, suspicions of electronic eavesdropping, shady characters, deceptions, confusions, embarrassments, and rumours of sinister threats – has been interpreted by some as evidence of naivety, gullibility and paranoia. But the first point to note is that Arnold himself provides this "evidence against interest" in what has every appearance of a full and frank memoir.[28]

Secondly, there is the fact that Arnold, and Capt. Smith – who at Arnold's invitation joined him in Tacoma after the first day – *were* the subject of attention from Army Counter-Intelligence at the time, and *were* also being deliberately hoaxed by a real-life shady character called Fred Crisman, who under his *alter ego* of Jon Gold was much later to become embroiled in the Kennedy assassination prosecution.[29] A few weeks after surviving an attempted murder by gunshot from a passing car outside Tacoma in 1968, Gold *aka* Crisman was summoned by New Orleans D.A. Jim Garrison to testify after being identified as one of the suspiciously clean-shaven and well-shod "train yard hobos", photographed leaving the area after the shooting, accompanied by "police" wearing radio earpieces.[30]

An ex-fighter pilot who claimed to have worked as a State Department special investigator during 1946-7, Crisman clearly exerted strong influence over his reluctant partner and probable dupe Harold Dahl. Arnold records that he and Capt. Smith were indeed very suspicious at times of the Maury Island story and especially of

Crisman, whom they caught in an apparent lie. But Dahl's innocent believability and the surrounding events kept them guessing.

Those events included warnings from United Press's Ted Morello and local reporter Paul Lance about possible threats to their safety, and the shocking deaths of Capt. Brown and Lt. Davidson in a plane crash *en route* back to Hamilton Field from their meetings with Arnold and Smith in Tacoma. Arnold cannot be held responsible for this concatenation of odd events, fuelled by press speculation. Throw into the mix the 'saucers', and the unsought celebrity, and it is little wonder that Arnold – in the thick of it – did not see all of these events quite as coldly as an uninvolved armchair sceptic with the accumulated hindsight of 60 years. Nevertheless, and despite the later influence of *Amazing Stories* publisher Ray Palmer, I find Arnold's account of these events[31] to be quite thoughtful; I would say even conscientious.

It would be fair to say that Arnold was caught up in the events in Tacoma, and perhaps took actions which at the time may have seemed sensible but which with many years' hindsight might be thought questionable. But the world had been caught up in the saucer mystery and had turned Arnold into Mr Flying Saucer. Why do we imagine that an event like this would not change the man? It did. We might be doubting Arnold's sincere conviction had it not. He was, of course, uniquely certain that his "saucers" did indeed exist, and suspected that they may have been a military secret, and this may have made him more than averagely inclined to wonder if the strange goings-on in Tacoma might be sinister; but as far as one can judge from his ingenuous account, and to the limited extent that it is possible to calibrate it from independent sources, it does not appear to me to have made him inclined to any more gullibility and paranoia than might have affected any of us in his shoes. Indeed, he records everything, including his own doubts, fears and mystification, with engaging frankness.

Years later Arnold did become, to some extent, an innocent victim of his own saucer mythology. But I don't see any evidence that on June 24 1947 he was a victim of anyone else's. Maury Island does not in my opinion cast a shadow back in time over June 24. These events play a part in the future unfolding of the saucer story, but that is a separate issue. I personally do not agree that they raise any serious questions about flaws in Arnold's character that would tend to undermine his original report relative to what we ought to expect from a witness of at least ordinary care and objectivity. The character testimony of those in a position to judge him at the time consistently portrays him as appearing conscientious, honest and not naturally very imaginative, and I tend to think that this profile fits what I see of the man in his own words and actions.

Arnold as 'Repeater' – His Subsequent Sightings

One very experienced researcher has suggested that "[Arnold's] credibility as a good observer is contradicted by the numerous UFO sightings he had in the weeks, months and years after his initial sighting. He became a 'repeater'."[32]

In the months and years after June 24 1947, whilst flying in the course of his business and/or during trips with a cine camera hoping for a repeat of his original sighting, Arnold did indeed have a number of other sightings of objects that puzzled him, and he filmed some. Why did he make these observations, and in particular what can they tell us about Arnold's mental set on June 24 1947? Do they tend to indicate that Arnold was a man primed by latent psychological factors to begin "seeing things" on June 24 1947? Or do they suggest that being fortuitously in the right place to see things on June 24 was what changed the man and primed him to begin looking for others? The state of the observing instrument (Kenneth Arnold) cannot be assumed to be a constant. We may need to recalibrate the baseline at June 24 1947.

To decide which of these scenarios fits the facts better it is important first to have some feeling for the nature of those later sightings. If they were highly exotic we might be tempted to conclude that Arnold's original sighting had been a trigger that allowed irrational thinking to flow to the surface, in the nature of a Damascene "conversion", a sudden unlocking of a dormant or latent tendency to fantasise.[33] But if the later sightings were not highly strange or exotic, and especially if it should appear the case that most or all of Arnold's reports contain the detail tending to permit plausible explanations of them, then this would favour the alternative that Arnold was simply sensitized by the original sighting to the possibility of seeing saucers.

Secondly, and semi-independently of the nature of each individual sighting, there is the matter of the frequency with which Arnold made them (relative to his opportunities and his efforts to see things). If the frequency suggests that Arnold was abnormally prone to make these sightings, or if the context suggests that he tended to see saucers with the least excuse, then we should not be able to rule out the interpretation that this reveals a pre-existing underlying credulousness on the part of Arnold as of June 24 1947. On the other hand, if the frequency seems unremarkable it would imply no such thing.

Arnold flew up to 100 hours a month on business at this period and carried a movie camera in the cockpit at all times after June 24, ever vigilant for a second sighting from the big-windowed cockpit of his CallAir mountain plane. Over the following 20 years or so he had in the region of 10 sightings in all, several supported with a movie record of some kind. Most of these sightings are trivial or explainable – either by Arnold's own admission or as suggested by information in his own accounts of them. The movie film clips showed nothing more than blurs and distant sun reflections, and none was worth anything as evidence, as Arnold himself and Ray Palmer both conceded.

Two weeks after his original sighting Arnold cabled a fellow pilot: "I flew over 1000 miles today (July 8, 1947) at 10,000 foot altitude hoping to get a moving picture but without success".[34] The previous day, July 07, Arnold had flown 7½ hours all over the Cascades area with Dave Johnson, aviation editor of the *Idaho Daily Statesman*, also without seeing anything.[35] Arnold continued seeing nothing at all that he thought worth an inch of film, not even a bird or a cloud or a distant plane, for over a month.

Then on July 29 when flying alone from Boise, Idaho to Pendleton, Oregon, descending into the La Grande valley at 5000 ft over the town of Union, Arnold found himself flying "head on" into a cluster of small "brass coloured objects that looked like ducks". Startled, he grabbed his camera and began filming. "Even though I thought they were ducks when I first saw them, I wasn't taking any chances". Arnold was heading roughly north. The objects, appearing a couple of feet across, veered to his right[36] past the plane, "fluttering and flashing a dull amber colour". They seemed to be round with a dark spot and looked "rough on top", but he couldn't be sure "because it all happened so suddenly". He "attempted to make a turn and follow them" but they disappeared to the east, apparently too fast for him to follow. The filming was "not very successful" and showed only a couple of tiny specks.

Many critics in later years[37] have argued that Arnold did see ducks, and that his considered opinion – he was "positive they were not birds" – damages his credibility as a witness. But this is an argument whose conclusion – that Arnold observed birds poorly – is inconsistent with its premise: that the clues indicating birds are "extraordinarily blatant" in his account.[38]

Arnold's *opinion* may be questioned, but when we ask, "what is the latent information Arnold provides?" we find that *despite* having an interest in persuading us of the strangeness of the sighting he describes the essential features of flying head-on at close range into a cluster of small objects that are the same order of size as ducks, which he initially thinks *are* ducks, that he says flutter like birds, and that fly in a cluster "like blackbirds". The head-on closure rate with a flock of ducks could have approached 200mph, meaning that they would have gone from dead-ahead at a range of 100m to falling behind the right wing in a little over 1 second, a very fast angular rate impossible to match in a turn and well fitting what Arnold described as "a terrific rate". They were lost to sight behind Arnold and "in the east", i.e., they were receding into the sun which was on the SE horizon. It was 5:30 am, about sunrise at Arnold's location. The reddened dawn light could also help to explain the "amber" colouration. Arnold said that he did turn and try to pursue, but they were too fast. He thought a lot about the possibility that they were birds, and assures us that he knew the wildlife in the area. But he also tells us: "I heard later that several farmers in the vicinity of Union had observed what they thought a peculiar cluster of birds that morning".[39]

It may be instructive to put ourselves in the position of a man who "knows" that there are saucers out there (because he has seen them, and because other respected authority-figures such as Capt. Smith have ratified his observation), who has been flying thousands of miles with camera in hand eagerly hoping to see more saucers, who knows that officialdom needs more to convince them to take him seriously, and who knows that the world's press and public are eager for more sightings – and then we should ask ourselves if we, with all these axes to grind, would have done so well in observing a confusing and fleeting encounter with a flock of unfamiliar birds, and described it so cautiously and objectively, including evidence 'against interest', in a book designed to definitively justify ourselves and our original sighting to the public.

Arnold snatched another short film of two objects that appeared to race under his plane "going to beat hell" near Mt. Lassen, California, on a sunny day in 1951. One of them appeared to "change density" and in illustration of this Arnold referred Dr James McDonald to another sighting by a Boeing test pilot who told him he saw an object like a manta ray whose wings "rippled" as it flew past his plane.[40] Arnold apparently saw a similar effect. This density-changing object seemed to be transparent or "wraithlike" and Arnold thought that he could see a pine-tree right through it as it passed between his plane and the ground, whilst its companion was "solid as a Detroit automobile".[41]

Once again the encounter was very fleeting. He shot some 16mm film, but the result was disappointing because, as he recalled later, he had trouble keeping the camera on the fast objects and flying the plane at the same time. He sent the film to Ray Palmer, who determined that something appeared on only 40 frames, or just a few seconds of film. Arnold told Dr Richard Read in 1965 that the film was "difficult to interpret"[42] and he may not even have seen it developed. He publicly asked Palmer about it in 1977. Palmer said that he had sent the film to the Air Force who replied that they could find nothing on it. Palmer claimed that it was returned with the 40 frames missing.[43]

So was Arnold startled by a low-flying fast jet near Mt Lassen? Did he see it accompanied by its shadow "rippling" over the forest landscape? From the very limited latent information in Arnold's description it seems possible, but we will apparently never know. We do know, however, that Arnold made no claims for the value of his film as evidence.

A third sighting, of a triangular object, was made by Arnold near Idaho Falls in July 1966. In this case not only did he make a film but there were many other witnesses. McDonald discussed the detail of this with Arnold by 'phone a few months later, and according to his notes Arnold told him that it had been explained in the papers as a possible pyramidal "weather balloon" or a high-altitude research balloon as apparently indicated by ESSA. Arnold collected many news clippings on the incident. Hundreds of people sighted the object moving SW apparently at great height, and a B-52 flew up to it at 54,000ft. It was "probably released from Minneapolis".[44]

In 1977 Arnold publicly discussed this sighting again and commented that on this occasion he'd obtained a "good" film, but only of what probably was a partly deflated balloon. I think it's interesting and characteristic – and we might argue that it is character-diagnostic – that when Arnold tells us about having one good film he openly says that it was probably a balloon.

Of other sightings made by Arnold over the years little or nothing is known.

Conclusion

So what should we say about the status of our "observing instrument", Kenneth Arnold, on June 24 1947 and in the following crucial days when its read-out was first obtained and studied? Do we have evidence that it was defective or maladjusted?

I think we have to say that we do not. Relative to the standards of caution and dependability of any ordinary observer in 1947, innocent of the sorts of exotic notions that have contaminated other observers since, Kenneth Arnold seems to have been, so far as we can tell, in fairly good working order.

That does not, of course, mean that his sighting report is simply a photograph of objective reality. What he saw remains uncertain and controversial. But what he *said* he saw appears to have been described in 1947 with rather good internal consistency, and we lack significant evidence that he did *not* see it as he described it. And whilst the origin of the "saucers" remains complex and intriguing, we can with some confidence put those bamboozling "boomerangs" back in the box of ufological canards they came from.

Martin Shough is a Research Associate for the National Aviation Reporting Center on Anomalous Phenomena (at the invitation of Dr Richard Haines, ex-NASA Human Factors head and NARCAP Chief Scientist) and is author or co-author of a number of papers and other publications. His main interest has been radar events. He has worked with several international collaborations to study cases in different parts of the world going back to 1946, and is the originator of a major ongoing effort to compile an historical catalogue of such events. He was trained in photographic and print science at London College of Printing and was a landscape artist and gallery-owner in the Highlands of Scotland for many years. He has also been a semi-professional folk musician.

In Search of the Last Mammoth

Seeking a living relic of the last Ice Age

by *Nick Redfern*

The mammoth: a majestic creature that roamed the lonely wilds of North America, the vast expanses of Western Europe, and the harsh lands of northern Russia during the Pleistocene era, and which is generally accepted as having become extinct somewhere around the end of the last Ice Age. Today, all that remains of this once mighty beast are a number of well-preserved carcasses found embedded in icy tombs, and the various bone and tusk fragments that still continue to surface from time to time. Could there, however, still be more – possibly even *much* more - just waiting to be uncovered?

For years, intriguing and sensational rumors have surfaced to the effect that in some of the more remote parts of our planet the mammoth might have existed until very recent times. While such a scenario is certainly controversial, and one that is completely derided by mainstream science and zoology, perhaps it is not entirely out of the question. For example, the related dwarf mammoth of Wrangel Island – located in the Arctic Ocean and situated between the Chukchi Sea and the East Siberian Sea – is known to have lived until approximately 1700 to 1500 BC, which is itself startling and highly illuminating.

Of course, many might say "absolute nonsense!" to the idea that the mammoth did not vanish in the time frame that science and mainstream zoology confidently assert. Others, however, just might be inclined to argue the point. For example, according to researcher Lindsay Selby:

> In the 1580s the Stroganoff family sent a band of Cossacks to hunt down a group of bandits in Siberia that had been stealing from their mines there. The leader of the expedition, Yermak Timofeyevitch, reported that beyond the Ural Mountains he met a 'large, hairy elephant.' The natives told him that the Kingdom of Sibir considered the giant animals a part of its wealth; they were valued as food and called 'mountains of meat'.

A Meeting of Monsters

Moving on, in 1873, the *Zoologist* published a sensational article titled "The Mammoth Still in the Land of the Living". Written by one Edward Newman, it stated:

> The *New York World* describes an interview between one of its correspondents and Cheriton Batchmatchnik, a Russian convict, lately pardoned by the government in consideration of his wonderful discoveries in Siberia, to which territory he had been banished for smuggling. Cheriton had escaped from the mines of Nartchinsk, and having reached the mountains struck southward

> for the Amoor River, intending to get to China. Meeting Cossacks he again turned north, and essayed what seemed to be a pass in the great Altai range.
>
> For thirty days he scrambled about seeking in vain for an outlet. At length he got out to the north, together with one of the branches of the Lena River. He then turned eastward and was entering the gorges of the Aldan Mountains when winter overtook him.

It was then, Newman said, that something truly startling occurred:

> Across the snow, came vast troops of animals, all going the same way. Cheriton followed them, arriving at the summit of the range, and there, far below his feet, landlocked between great ranges of hills on all sides, lay a valley fifty miles wide by one hundred and fifty long, and centered by a blue lake. He descended and found the valley warm and fertile, and full of animals. At night he made a fire and lay down beside it to sleep.
>
> During the night dark shadowy forms came over the water, splashing towards him, and seemed to seek what his fire might mean. The trampling of great beasts that crushed the willow-stalks like pipe-stems, on their way to the water's edge and that came and stood over him, breathing heavily and slow as they seemed to gaze at the fire with stupid wonder, made him afraid each moment of being overrun.

In the morning the fearful convict went in search of a safer camp site to spend the next night. Recrossing the meadow, Cheriton followed the edge of the slope until he came across the wide entrance to a cave. Noticing the rocky pavement leading into the cave was worn with trackmarks, he entered with some trepidation, when suddenly he heard…

> …a heavy startling snort, and there in the half light of the cave, standing alive, chewing the cud, and waving its proboscis to and fro, with a slow, majestic, gentle motion, he saw – a mammoth!

Continuing his own account of what reportedly took place, Newman reported that Cheriton described the mammoth as being "a very imposing looking animal, covered with reddish brown wool and long black hair.…being about twelve feet high, and eighteen feet long, with tusks projecting about four feet, and being eight to ten feet counting the curve." Nocturnal in its habits, and frequenting caves or the forest depths by day, the mammoth's skin was "bare on the upper surface of the ears, on the knees, and rump", and was "of a mouse-colour, and seemed very thick". Newman also noted that Cheriton did not see any young mammoths, just older animals that were "very peaceable animals, torpid and sluggish as old oxen",all of which paid little attention to the Russian convict in their midst.

A Battle of the Beasts

Things then got, frankly, somewhat unbelievable when, allegedly, yet another monstrous beast surfaced on the scene – this one being a classic lake-monster that supposedly rose from the depths of the aforementioned blue lake and engaged one of the mammoths in battle! Newman described the suspicious-sounding mammoth vs. leviathan saga thus:

> The lake was inhabited by a monster of which Cheriton was in constant dread, a sort of saurophidian, which he described as being thirty feet long, and armed with scales and horrible fangs. This monster – he never saw but the one – was master of the lake, and lived by devouring the animals which came by night to its brink to drink.

> Cheriton gives a graphic and exciting description of a contest which he witnessed one morning at early dawn between this crocodile-serpent and one of the mastodons. The battle which lasted more than an hour, ended in the discomfiture of the mammoth, which could barely limp away after having been constricted in the serpent's folds.

Newman noted that a month after the story first surfaced there was still no "official contradiction" of it and its extraordinary contents. Rather, Newman added, Cheriton's account had received positive commentary from a Professor Feuchtwanger before the American Association for the Advancement of Science.

According to the professor:

> The discovery of the mammoths in Siberia in the deep gorges of the mountains near the Lena Viner, which was lately published having been made by a scientific Russian convict, who had seen five living animals, twelve feet in height and eighteen feet in length, with projecting tusks four feet long, excites some discussion in Europe.
>
> I think it worthy of inquiry whether the mammoth of the post-tertiary period, discovered during this century in Siberia, near the same river, can have any relation to the convict's discovery. Thousands of these animals have been found buried in the ice, with their well preserved skins, and thousand of tusks are brought to England this day for the use of the turner. These are of nearly the same dimensions as those seen by the Russian. The convict has received an unconditional pardon, on the recommendation of scientific men who have investigated his statements and believe them to be true.

Not everyone is quite so sure that the story has merit, however.

A Tall Tale

Fortean researcher and writer Chad Arment says: "Occasionally, in discussions of the possibility of living mammoths, there will be mention of a sighting in Siberia that was published in 1873. The story itself was written for the *New York World*, recognized even then as a source for a number of tall tales and newspaper yarns."

In support of this statement, Arment cited the following from the August 23, 1873 edition of Cincinnati's *Daily Gazette*:

> The *New York World* is fond of remarkable sensations, especially those of an apparently scientific nature. Not long ago, it told, on the pretended authority of a Russian fugitive from Siberia, of the discovery of mammoths and other extinct animals among the Siberian Mountains. It once described the transference of one man's brains to another's skull so that the reconstituted individual walked about in his own flesh but with a new mentality. Its wonderful interviews with Bismarck and Eugenie, which were too good or too bad to be true, will be remembered by our readers. Its latest attempt is a report of the discovery of the famous library said to have been carried by Tamerlane to Samarcand. The account goes on to describe some of the literary treasures of antiquity included in the collection. An air of truthfulness is given to the story from the fact the Russians have laid open the surrounding Asiatic districts, hitherto almost sealed to Europeans. It happens, however, that Arminius Vambery, the enterprising traveler who visited Samarcand in the disguise of a dervish a few years ago, made particular inquiries, but could hear nothing of it. If it had been in existence, the Russian commanders would, months since, have made the fact known to the world. The *World's* fiction is so pleasant that we could wish it were fact.

Of course, the undeniable fact that Newman's story is still told today demonstrates one of the premier problems that besets much of Fortean research: namely, that while it contains enough nuggets of data to deem it worthy of study, the insertion of a *Jurassic Park*-like monster that does battle with a mighty mammoth inevitably results in the rolling of eyes and hoots of derision.

Just like Ray Santilli's notorious "Alien Autopsy" film; the 1967 footage of Roger Patterson said to show nothing less than an honest-to-goodness Bigfoot; and the tales of George Adamski, the strange saga of Cheriton Batchmatchnik continues to provoke keen debate. The believers continue to believe and, as for the disbelievers: well, they continue to disbelieve! Thus it has always been within Forteana, and doubtless it always will be.

The Reports Continue

Nevertheless, accounts of still-surviving mammoths continued to surface. In the late 19th century, for example, researcher Bengt Sjorgen learned that tales were circulating widely (and wildly) in remote parts of Alaska about giant, hairy tusked creatures that lived deep under cover of the huge, ancient forests. Such reports of the "hairy elephants" in question extended to equally wild parts of both Canada and Siberia.

Similarly, in February 1888, the New Zealand-based *Argus* newspaper reported on the apparent discovery in Alaska of strange tracks that had been found by the Stick Indians in the vicinity of the White River. An *Argus* journalist stated:

> One of the Indians said that while hunting, he came across an immense track sunk several inches in the moss and larger around than a barrel. The Indian followed up the curious trail, and at last came in full view of his game. These Indians as a class are the bravest

> of hunters, but the immense proportions of this new kind of game filled the hunter with fear, and he took to swift and immediate flight. He described it as being larger than the post trader's store, with great shining, yellowish tusks, and a mouth large enough to swallow him at a single gulp.

Then, during October 1899, the story of one Henry Tukeman surfaced – a man who claimed to have killed a mammoth that was subsequently donated to the Smithsonian Institute, Washington, D.C. The tale was denounced as nothing more than a sensationalized hoax (which it probably was); however, some researchers still cling on to the belief that it just might have a grain of truth to it, and that the hoax angle was possibly introduced to try and lay the controversy firmly to rest.

And also in the late 1800s, several reports of "large, shaggy beasts" were given to Russian authorities by Siberian tribesmen, but no proof was ever forthcoming – some might say inevitably, and conveniently.

Twentieth Century Survivors

Moving on, Lindsay Selby has pointed out a more 'recent' mammoth tale involving a Russian hunter who claimed to have seen a pair of the beasts in 1918. The story was recorded by M.L. Gallon, the man in charge of the French Consulate in Vladivostok. According to Selby:

> Gallon claimed that the hunter did not understand what he meant when he referred to the beast as a mammoth, but the hunter simply maintained that it was similar to pictures of elephants he had seen. Although Gallon shared the story with friends when he returned to France later that year, he was not persuaded to publish the account until 1946.

The astounding story was originally told by the legendary cryptozoologist Bernard Heuvelmans, who quoted the hunter thus:

> In the second year I was exploring the *taiga*, I was very much struck to notice the tracks of a huge animal. I say huge tracks, for they were a long way larger than any of those I had often seen of animals I knew well. It was autumn. There had been a few big snowstorms, followed by heavy rain. It wasn't freezing yet, the snow had melted, and there were thick layers of mud in the clearings. It was in one of these big clearings, partly taken up by a lake, that I was staggered to see huge footprints pressed deep into the mud. It must have been 70 cm across the widest part and 50 cm the other way, so the spoor wasn't round but oval.

Following the tracks, the hunter came across a massive heap of dung composed of vegetable matter. And where the tracks entered into the forest, the branches of trees were broken some 10 feet up. Intrigued, the hunter followed the track for some days…

> …thinking that perhaps I should never see them, and also a bit afraid, for indeed I didn't feel I was big enough to face such beasts alone. One afternoon it was clear enough from the tracks that the animals weren't far off. The wind was in my face, which was good for approaching them without them knowing I was there.
>
> All of a sudden I saw one of the animals quite clearly, and now I must admit I really was afraid. It had stopped among some young saplings. It was a huge elephant with big white tusks, very curved; it was a dark chestnut color as far as I could see. It had fairly long hair on the hindquarters but it seemed much shorter on the front. I must say I had no idea that there were such big elephants. It had huge legs and moved very slowly. I've only seen elephants in pictures, but I must say that even from this distance (we were 3000

> m apart) I could never have believed any beast could be so big. The second beast was around; I saw it only a few times among the trees. It seemed to be the same size."

And the tales do not stop there. A further sighting allegedly occurred during the Second World War when a Soviet Air Force pilot reported seeing a small herd of such creatures while he was flying over the frozen wastelands of Siberia. And although this particular report is highly intriguing – in the sense that it suggests the mammoth may have been a resident of this planet up until extremely recent times – it frustrates researchers by virtue of the fact that trying to chase the story to its original source has not yet proven successful.

Spectral Beasts

The story is not quite at an end, however. There is yet another aspect to this controversial saga – one might even call it a downright spooky aspect, too. It came to me on January 2, 2009 from a woman named Jill O'Brien and is, to this day, one of the oddest tales to ever cross my path. And, believe me: my path has crossed with many a weird tale in its time! According to Jill, on September 1, 2008, she came face-to-face with a baby mammoth in Alaska's Wrangell-St. Elias National Park and Preserve. And, here I was thinking that Sarah Palin was the most monstrous and diabolical thing ever to surface out of Alaska!

Well, this was not a case I could afford to ignore – particularly as a result of the fact that, at the time, Jill was living in Oklahoma City, which is only a couple of hours' drive from my Arlington, Texas home. That's right: it was road-trip time. And so, I headed northwards to chat with one of the very few people to claim a sighting of a still-living Mammoth. But, as will shortly become clear, things weren't quite so...well...clear.

I set off early on the cold morning of January 9, and by lunchtime was sitting opposite Jill, in the small, rented apartment that was to be her home before moving on to pastures new in Nebraska in February. Despite the fact that the skeptics and naysayers love to portray those who claim encounters of the truly unusual kind to be slightly off-kilter, such a claim could not be made about Jill: very down-to-earth, lucid and intelligent, she worked for a well-known bookstore chain and related her remarkable story in a coherent and clear fashion.

Established in 1980 by the Alaska National Interest Lands Conservation Act, Wrangell-St. Elias National Park is the largest U.S. national park, and covers no less than 20,587 square-miles – which, trivia fans may be interested to know, is greater than the land mass of no less than nine of the American states. It was very near to the base of the 18,009-foot-tall Mount St. Elias – which lies on the Yukon-Alaskan border – that Jill's encounter took place.

Jill told me that she had a particular passion for photographing mountains and mountain-ranges; and Mount St. Elias was most certainly one that she wished to preserve for posterity in her ever-growing collection. She was all set up to take a "great shot" of the mountain, when to her left she heard "a funny crunching and thudding on the ground." She turned her head to where the sound was coming from, and was "totally freaked" to see a small Mammoth "maybe four feet [in height]" race past her, glance quickly in her direction as it did so, and then vanish into what resembled a "small, black cloud of smoke that kind of sucked into itself and was gone."

Jill fully conceded that the event – which lasted perhaps no more than seven seconds – stretched credulity to its absolute limit; but, as she told me in tones that came across as being wholly earnest in nature: "It was a mammoth, a baby mammoth. It was there one second; it looked at me, and it was gone: like it went invisible or just vanished. But that is what happened. Sorry, I know it's probably not going to interest you."

Needless to say, Jill's case most assuredly did interest me. Could it be that Jill really did cross paths with a mammoth – albeit the ghostly remnants of one that shuffled off this mortal coil thousands of years ago? Of course, some might dismiss such a story as the ravings of a fantasist or mistaken identity. But I like to think not. Rather, I prefer to imagine that while the mammoth is almost certainly no longer with us today, it may very well have survived and thrived for much longer than many have imagined possible. And, perhaps Jill O'Brien's encounter at Alaska's Wrangell-St. Elias National Park and Preserve is evidence that even if, in its physical form, the mammoth is a thing of the past (and perhaps the very recent past), it hasn't yet quite given up the ghost – so to speak!

Musing on the Mammoth

So, did the mammoth exist for longer than science suggests? Could it, even, still be with us? Or are all of the tales merely friend-of-a-friend accounts, myths, hoaxes, and misidentifications? Personally, I have been fascinated – and perhaps, I'll be the first to admit, slightly obsessed – by these stories for years. And, just like *The X-Files* said: I want to believe. I really do. Of course, the skeptical and logical parts of my brain tell me that the mammoth is an utterly dead creature; one whose existence came to a tragic and definitive end thousands of years ago – aside from in the *Mammoth* movie of 2006, and in the 2008 Hollywood production *10,000 BC*, it could be argued.

But who can deny the appeal that stories like those I have cited above create in the minds of thrill-seekers everywhere? I certainly can't, that's for sure. If there was one expedition I could go on, and if both funding and time were utterly unlimited, it would be to the old stomping grounds of the mammoth. Okay, I know full well that the chances of actually finding a living, breathing mammoth are not just miniscule; they're beyond miniscule. But, as long as there is even just the remotest of possibilities, I know I'll never be truly satisfied until I go and seek out the creature for myself. Until that day hopefully comes, like everyone else I can only continue to wonder if, far from prying eyes, small pockets of mammoth continue to walk the earth; utterly oblivious to the controversy they are currently creating in the pages of this very volume of *Darklore.*

Nick Redfern is the author of many books on the world of the paranormal, including *There's Something in the Woods*; *Memoirs of a Monster Hunter*; *Final Events*; *Contactees*; and *Monsters of Texas* (the latter co-written with Ken Gerhard). Nick writes regularly for *Fortean Times*, *UFO Magazine*, *Paranormal Magazine*, and *Fate*. He lives in Arlington, Texas, USA and can be contacted at www.nickredfern.com

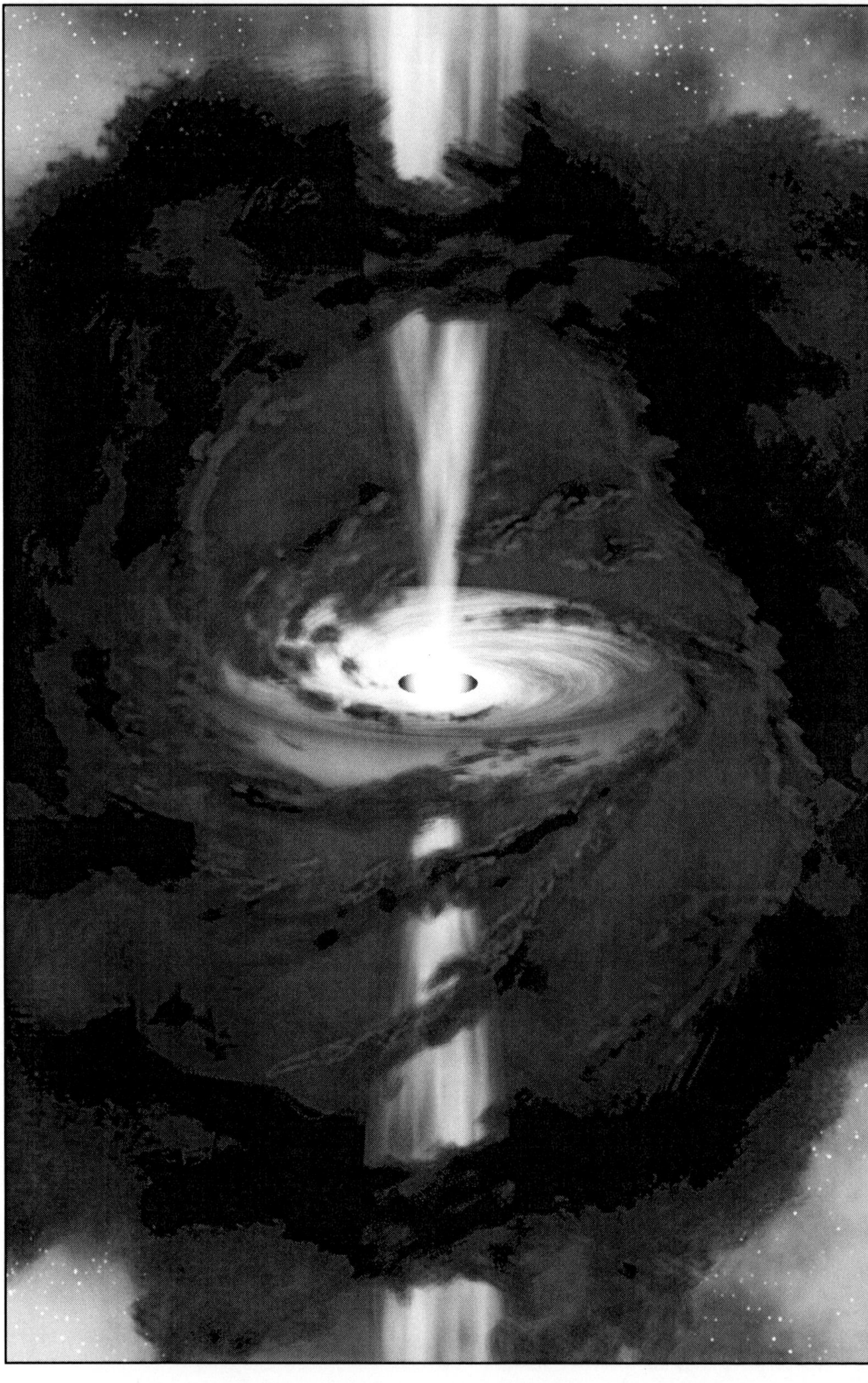

Dark Seed

The ABCs of the Sirius Mystery

by ***Blair MacKenzie Blake***

I do not know if it was the 23rd, but on one hot sunny day in July, back in the early 1980s, I was surprised to see a copy of a book entitled *The Sirius Mystery* lying on the front doorstep of my home in Los Angeles. Although there wasn't any note attached to the battered hardcover, I was quite excited to finally be holding that which I was having such a difficult time finding. For nearly a year I had been trying to obtain the book through the public library system, only to be informed at each branch that their only copy was missing. In some cases the book had been checked out and never returned, while in others it had been stolen for what one librarian suspected to be "esoteric purposes." Over the next few months, I treasured the book as if it was the only copy in existence. I pondered its abstruse content and pursued further areas of research. Eventually I placed it on a bookshelf to gather dust. Who ever had left it on my doorstep is still unknown to me today, but, over a quarter of a century later, with glossy reprints now readily available, what about the Sirius mystery itself? Have new findings in science and elsewhere served to deepen the mystery posed by the book, or does certain information recently called into question warrant the entire matter to be dismissed as sheer fantasy?

The Secret Traditions of the Dogon

First published in 1976, *The Sirius Mystery* by Robert K.G. Temple attempts to explain how certain high-priests of the Dogon – a supposedly primitive tribe living along the Cliffs of Bandiagara on an escarpment just south of Timbuktu in the present Republic of Mali, West Africa – came into possession of detailed and accurate information concerning the binary (or possibly trinary) Sirius star system. This specific knowledge includes astronomical peculiarities that the tribal elders have allegedly preserved for at least 800 years.

For those who examine Temple's work without prejudice, the secret traditions of the Dogon and neighboring tribes is startling to say the least. Rather than venerate the brightest star in the night sky, Sirius (relatively close to earth at 8.7 light years away), the Dogon attach greater significance in their rich cosmology to its dark companion, Sirius B,[1] a star that is totally invisible to the naked eye. Not only do the Dogon priests seem to be aware of the invisible companion's existence, it would appear that they also know that it is an exotic type of star known as a white dwarf, claiming it to be one of the heaviest stars in the universe, a fact that astronomers now know to be true. And if that were not enough, they also know that the trajectory of its orbit is elliptical, taking 49.9 years to complete around Sirius A, an event that they celebrate with an elaborate ceremony. It has also been claimed that Dogon sand diagrams even show this elliptical path with Sirius A positioned off-center, or at a particular focus of the ellipse, which would also be astrophysically accurate. Could this all be due to chance, or did someone (or something) provide these isolated sub-Saharan peoples with information that is relatively new to the rest of the modern world?

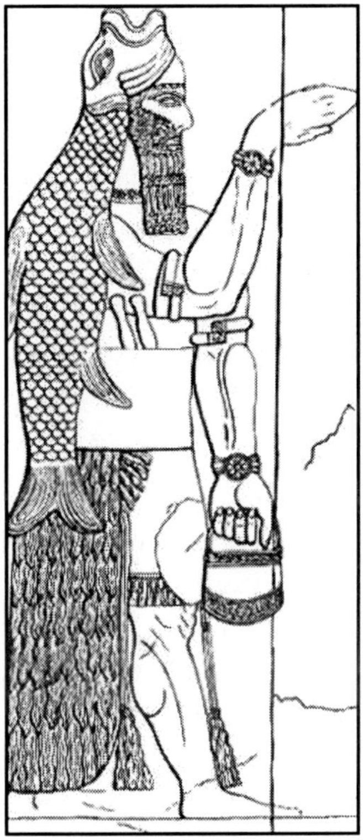

Using fragments of ancient Sumerian, Babylonian and Egyptian 'mythology' to solve the puzzle, Temple traces the origins of the Dogon's inexplicable knowledge to an advanced race of aquatic extraterrestrial beings called the 'Nommo' that inhabit a planet in the Sirius system, and who visited the earth around 5000 years ago. The tradition of the dark partner of Sirius and of the emissaries of the god 'Amma', who bequeathed civilization to those who they came into contact with, was preserved for centuries among the initiated, handed down from generation to generation, until it was finally revealed to two reputable French anthropologists in the 1940s. The anthropologists, Marcel Griaule and Germaine Dieterlen, who had lived among the Dogon for years and apparently gained their utmost trust, subsequently published their innermost teachings, and it was these articles that piqued Temple's interest, prompting him to delve further into the matter.

Taken More Seriously

Owing to his extensive research and the perceived scholarly nature of the book, many readers found the evidence convincing enough to support Temple's paleocontact hypothesis. Unlike the much-maligned work of von Däniken and his ilk – whose researches into ancient astronaut theories are severely flawed by factual inaccuracies – Temple's postulation of galactic visitations being responsible for jump-starting civilization generated a lot of excitement. However, despite the apparent intellectual integrity, the author still had more than his fair share of critics, and the book was met with fierce opposition from those not inclined to stray too far from the prevailing academic view of such things. For example, archaeoastronomer Dr. E.C. Krupp suggests that the Dogon sand drawings could be interpreted as "a sexual mnemonic that equates agriculture with human procreation", rather than representing the peculiar elliptical orbit of an 'invisible' star. Others who believed that the Dogon's anachronistic knowledge wasn't just a string of coincidences thought that their star lore was derived from early Christian missionaries who (for some strange reason) felt it necessary to teach a near-Stone Age African tribe the celestial mechanics of a stellar remnant white dwarf.

If the idea of outside interference isn't a satisfactory explanation, then perhaps the simplest solution is that the anthropologists themselves hoaxed the entire thing. Although Sirius B wasn't first photographed until 1970, its presence as an obscure companion of Sirius was guessed at back in the mid-19th century by the German astronomer Friedrich Wilhelm Bessel after observing tiny irregularities (wobbles) in the slow proper motion of the larger star over a ten year period. His suspicions were finally confirmed in 1862 when the dark star was first observed by Alvan Graham Clark using what was then the largest refracting optical instrument in the world. Therefore, it was certainly conceivable that members of the scientific community were aware of the unusual characteristics of Sirius B at the time that the anthropologists were studying the Dogon's beliefs, and history has shown that even 'reputable' members of academia have been known at times to take certain matters into their own hands.

Other critics of Temple's extraterrestrial conclusions have claimed that Marcel Griaule over-interpreted the Dogon high priests' answers to questions concerning their cosmological beliefs and calendrical mythology, suggesting that he was basically told what he wanted to hear. This is just short of saying that the eminent anthropologist put the words into the Dogon elder's mouth, thus fabricating (even if inadvertently) a great mystery for possible academic glorification. But as Temple has pointed out, the problem with this is that in his report Griaule didn't trumpet the Dogon's apparent knowledge of the unseen companion. Instead, showing admirable restraint, he merely placed it as a footnote that stated, "The problem of knowing how, with no instruments at their disposal, men could know the movements and certain characteristics of virtually invisible stars has not been settled, nor even posed." Although this would seem to confirm that Griaule had an interest in astrophysics – a fact some critics have used to bolster their case that he was the sole perpetrator of the hoax – the other side of the coin is that he used the words "virtually invisible", when Sirius B is *totally* invisible to the unaided eye. Though it is true that we have to take his word for the tribe's most secret traditions – Griaule being the only outsider who managed to gain their confidence – it doesn't seem fair to accuse him of hoaxing the mystery simply on the grounds that other researchers haven't been able to find other modern day Dogon informants who are aware of the same knowledge of a dark companion star, let alone a third stellar component (see below). After all, most esoteric traditions are reserved only for the initiated.

In a much-expanded and revised edition of *The Sirius Mystery* published in 1998, Temple wastes little time answering (indirectly) certain of his critics by announcing, in dramatic fashion, the recent discovery of Sirius C, a small red dwarf star whose existence was 'predicted' by the Dogon. Temple's evidence for Sirius C is based in part on an article published in 1995 by French astronomers Daniel Benest and J.L. Duvent, who, after examining numerous reported observations of a third body in the 1920s, suggested that a six year perturbation in the motion of Sirius A-B *might* indicate the existence of Sirius C. However, in spite of earlier claims by astronomers and the intriguing irregularities in the motion of Sirius A-B, today this additional partner is still the subject of controversy, and without any observational evidence, hardly grounds for any vindication on Temple's part.

In the revised edition Temple also indulges in numerical mysticism such as sacred ratios in harmony theory (The Comma of Pythagoras, etc.) and its possible relation to esoteric astronomical lore. At its simplest, this could be described as the 'music of the spheres' beyond our solar system, but, in particular, the direct correlation of specific numerical values of the mass ratio between Sirius B and our own sun. Because both stellar bodies inhabit the same cosmic 'cell', according to Temple a type of "instantaneous communication" could exist, as if certain elements were not separated by temporal distance. Temple calls this self-organizing, possibly living entity "The Anubis Cell."

Although numerical values based on more recent (space-based) observations (as well as others yet to be determined) would appear to significantly alter his findings (such as the numerical coincidences between the Giza pyramid complex and astrophysical measurements involving the Sirius system), the notion of harmonic resonance and mutually-shared properties between neighboring solar systems might be worth further examination. Especially when it comes to the 'Starseed Transmissions' of Dr. Timothy Leary and others. And though Temple is quick to warn us that this interstellar telepathy is likely to be exploited by "every nut in California", ironically such channeled inspiration from Sirius might have been responsible for his receiving the information pertaining to the Dogon that initially sparked his interest in the mystery, and which ultimately led to the publication of his highly influential book. (More about this later.)

Stargate Conspiracy?

Related to the idea of privileged communications or even nonmaterial interaction of souls via the Anubis Cell, another popular book about Sirius-related beliefs theorizes how certain revolutionary discoveries about the nine principal gods (the Great Ennead) of ancient Heliopolis made by scientists, scholars and other powerful individuals "tie into a dangerous conspiracy nearly fifty years in the making." First published in 1999, *The Stargate Conspiracy* unveils strange connections between the 'new Egyptology', esoteric Freemasonry, occult fraternities and a plot by an unknown agency, using mystical transmissions supposedly channeled from intelligences from the Sirian comity, for some nebulous socio-political agenda. According to the authors, an important figure "where the genesis of the Stargate Conspiracy is concerned", is William Dudley Pelley, author of a very rare book entitled *Star Guests: Design for Mortality* (1950). Believing it to be the first collection of channeled material from extraterrestrials ever published, once they managed to obtain a copy, the authors were "staggered to find that it brings the whole story round full circle, as it focuses not just on Sirius, but actually on Sirius B…". Actually, what the book of 245 pages mentions on two occasions is: "a mammoth planet revolving about Sirius as its sun, completing its stupendous orbit in something like 49 of our solar years", and perhaps even more telling, "the question as to whether Sirius is attended by a satellite or accompanied by a companion seems still to be disputed." More than likely, rather than channeling godlike entities from some fantastic Sirian planet, the author got these ideas from reading a copy of Richard Hinckley Allen's *Star Names and Their Meanings* (1899), which evidently could even be obtained in the Federal prison system in the 1940s. What awaits those patient enough to examine the "automatic writing" in *Star Guests* is a nauseating string of platitudes, typical 1950s flying saucer contactee drivel, pseudo-scientific zingers, and new age 'soulcraft' doctrine with a Biblical slant that one might expect from the white supremacist founder of the fascist "Silver Shirts of America" (but hopefully not from beings from sublime dimensions controlling a miraculous pencil).

Sirian Migration, DNA, and Shamanic Flight

With ornate prose like "even with its influx of interstellar spirits migrating to our terrestrial ball as the progenitors of shapes hideous to think they might make in diabolical caprice", Pelley's book could easily have served as a template of sorts for today's proponents of ancient astronauts being responsible for human creation through genetic engineering. Using Griaule's anthropological report *Conversations with Ogotemmeli* (1948) as the source for information concerning Dogon religious ideas, in *The Master of Speech* author Shannon Dorey borrows Pelley's concept of Sirian migration, with the advanced cosmology of the Dogon analyzed in great detail. As where Temple hypothesizes that the alien visitations were responsible for the civilizing of humankind, for Ms. Dory the utter simplicity of the Dogon's oral history and drawings (which survived in its purest state due to the tribe's isolation) ingeniously preserves the truth about human existence. Dogon iconography and symbology – such as the granary, smithy, fibrous skirts, storeyed house, jackals and anthills – actually represent complex scientific procedures relating to DNA sequencing that enabled the aquatic and androgynous Nommos to produce biologically engineered hybrids tailored to their exact specifications.

Of course it's up to individual interpretation as to whether or not blacksmiths descending to the earth on rainbows symbolically represent actual alien spacecraft that also served as laboratories for disturbing genetic experiments necessary for the survival of their species. Yet, if we suppose that the advanced cosmology and other enigmatic beliefs of the Dogon are a modern remnant of an exceedingly ancient linage, could there be a more plausible explanation for the source of this deliberate instruction? Perhaps lost in the glare of thunderous B-movie rocket technology is the true (or, at least, partial) identity of this knowledgeable godlike teacher – an intelligently designed technology previously wholly unsuspected that not only imparts the rudiments of civilization, but also the latest findings in contemporary physics. Might this idea-transference be something akin to a chemical stargate, as proposed by Picknett and Prince, authors of *The Stargate Conspiracy*?

Drawing from the pioneering research of Swiss anthropologist Jeremy Narby in his book *The Cosmic Serpent, DNA and the Origins of Knowledge* (1995), the 'Stargate' authors argued that the real source of unexplainable knowledge responsible for great achievements in civilization are messages encoded in DNA, which itself possesses a consciousness that can be contacted by those who tune their own consciousness to the right frequency. Using the serpentine imagery of the DNA double helix, the specific knowledge of the cosmic serpent-teacher – such as generous hints in astronomy about the Sirius star system – might be accessed through the exalted states of consciousness experienced by shamans.[2]

However, when Narby speculates that these messages imparted to the shaman by guiding spirits may have been placed in DNA by "entities that were rather clever", we find ourselves again having to consider the genetic experiments performed by the Nommos. Yet, according to Narby and others, it's even possible that DNA serves as a *receiver* of information from a nonlocal domain, once again bringing to mind Temple's "Anubis Field." But with regards to the possibility of a chemical stargate, there is one publication I know of concerning a shamanic aspect to the Sirius mystery that is sure to be of interest.

The Fire Trail of Oge-Make

The January 1948 issue of the pulp magazine *Amazing Stories* contains a curious little tale involving a sacred Navajo ceremony that shares certain features related to the advanced cosmological knowledge now associated with the Dogon tribe. Unlike Pelley's vague allusions to Sirius B as the "seat of the Godhead", in the *Amazing Stories* tale, during the mystic adventure when the "pale son" Oge-Make ascended on a path of flames, traveling through both time and space, he returned with an incredible story involving "Sirius and its dwarf companion." He also described Sirius B as a small sun with a heavy core circling a larger neighbor. Besides these particulars, of further interest for fans of Temple's *The Sirius Mystery* is Oge-Make's account of the denizens who inhabit a planet in the Sirius system. These creatures were said to have a certain fish-like quality, living in a watery medium in which they were propelled by tail-like appendages. Compare this to the ancestral spirits worshipped by the Dogon – the Nommos, who were required to live in a watery environment whilst on the earth, thus mimicking their home world.[3]

As for Oge-Make's journey into the star-dusted blackness, besides the hypnotic rhythm of native drums and the throbbing chant by firelight, we can assume that the ancient ceremony also involved the consuming of a certain plant with psychoactive properties. His drinking of an unpleasant tasting concoction prepared by the Medicine man is followed by a sudden onslaught of vivid colors while staring into flaming cedar boughs. Once transformed into a higher body of light, he is able to communicate with the spirit guide, who is endowed with a similar radiant vehicle. And it is while in this altered state of consciousness, using what today might be called 'remote viewing', that he is given a tour of the Sirian "Planet One."

Quickly we learn that all is not well in the paradise of "Planet One." In the year 50,050 of their time, which on Earth time was "long millenniums before the memory of man", the core of the

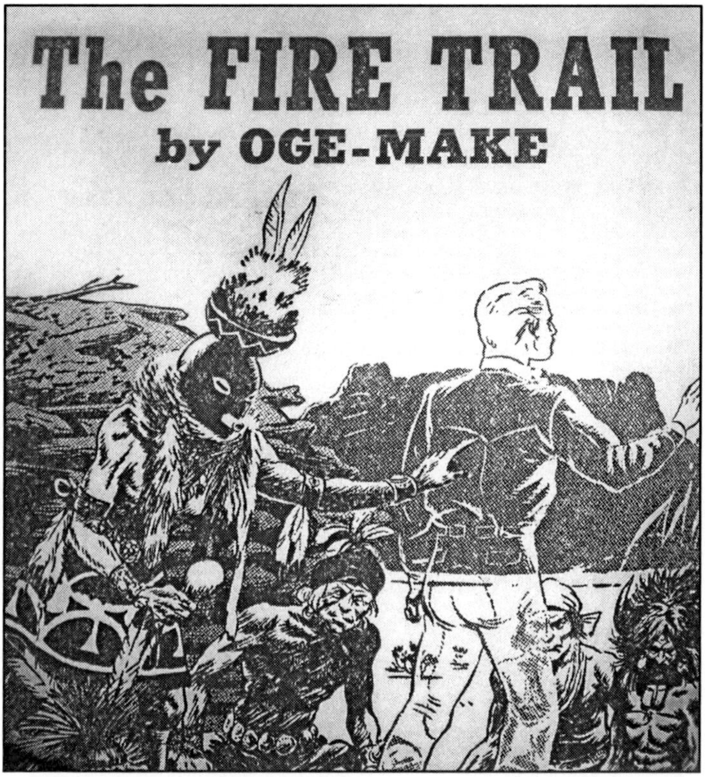

"The Fire Trail" illustration from *Amazing Stories*

planet had become heavier through thousands of years of atomic disintegration. Freaks of nature were developing due to deadly radiations, and it was their hope to find another planet where the least infected progeny could survive. While the contaminated planet's leaders debate whether or not they are justified in taking another planet in a distant star system by force, something goes seriously wrong with their own experiments with a futuristic weapon, causing their entire planet to catch on fire like "a picture soaked in gasoline, to which someone had touched with a match."

As a result, "Planet One" becomes a sun – that which is later revealed to be the collapsing white dwarf known to us today as Sirius B.

Amazing Stories

So what are we to make of this strange little tale involving a chemical stargate that enables one access to worlds beyond the spectrum of tangibility? Right from the start, the editor of *Amazing Stories*, Raymond A. Palmer, tells his readers that the January 1948 publication (Vol. 1, No 22) is not an ordinary issue. This is because they are running a story that he has no hesitation calling "TRUE". In a side box to the actual story Palmer again stresses the validity of the tale, claiming that not only is it based on an accurate account of what occurred in the summer of 1947 during a special ceremony held by Navajo, Hopi, and other Pueblo Indians, but that it contains a message of vital importance. However, considering Palmer's rather dubious reputation – he has been described as "the trickster of the pulps" and "the P.T. Barnum of ufology" – we probably shouldn't trust him any more than your average carnival barker. Even so, as we shall soon discover, many major elements of *The Stargate Conspiracy* come into play. First of all, it is almost certain that the author of the story, Oge-Make, was actually the science fiction writer L. Taylor Hansen, who contributed other material for the alternate reality magazine in the late 1940s. In fact, as a footnote to the story, Palmer admits as much, stating that a mutual friend passed on the tale to him. This was the "scientist" L. Taylor Hansen, who considered it to be interesting material for the magazine's readers. As for the author herself, it's hard to find out much information about Hansen, other than she wrote a book in 1963 entitled *He Walked The Americas*. This book focuses on various Native American Indian legends concerning a wandering prophet who spread the ancient wisdom of Atlantis and other mythic realms of a past golden age in preparation for a new global religion.

Although the 'scholarly' author describes herself as having a Masters Degree in both archaeology and anthropology from Stanford University (as well as being an expert Egyptologist!), no one has yet found any verification of her even being a Stanford alumnus. But regardless of her credentials, the fact that she claimed to have these degrees from Stanford may be telling in that the 'Stargate' authors consider the Stanford Research Institute (SRI), founded by Stanford University in 1946, to be an important link to the conspiracy.

According to Picknett and Prince, SRI is now well known for playing a key role in the new Egyptology, using both remote sensing and remote viewing techniques while searching for a possible ancient stargate technology at Giza. As is explained in their book, this stargate offers a means of transporting a person to other worlds, with the method of looking for it (i.e. remote viewing) involving an out-of-body experience in which the viewer is mentally able to travel across both time and space, and then report back what was seen. Sound familiar?

If the 'Stargate' authors are correct in their assumption of a dangerous conspiracy, in light of certain revelations imparted by the mysterious author of "The Fire Trail", it would appear that certain of the 'conspirators' have been around since at least as far back as 1948.[4]

Blue Star Kachina

Given the author's interest in Native American culture, the tale involving Oge-Make's dire cosmic warnings may have been based on the Hopi prophecies of the Blue Star Kachina, Saquasohuh. Associated with Sirius and "those that came here in the beginning", according to current Hopi belief, "though it is far off and still invisible, it [Saquasohuh] will make its appearance soon." The Hopi elders also claim that when the Blue Star Kachina dances in the plaza and removes its mask in front of the uninitiated children (i.e. the general populace), the "Day of Purification" will begin. Hence the appearance of this 'star' marks the destruction of the world as we know it. As it turns out, the Dogon tribal priests have a similar belief. In Griaule's anthropological journal *Le Renard Pâle* (1965), they maintain that the Nommos will come again: "A certain 'star' in the sky will appear once more, and will be the testament to the Nommos' resurrection."

The Bestirring of the Nommos

Concerning the return of our mythical ancestors, in the revised version of *The Sirius Mystery*, Temple suggests that perhaps the ancient visitors never returned home. And if they didn't make the journey back, he further speculates their space "ark" might be parked in orbit in one of the outer planets in the solar system. He even proposes, in all seriousness, that this immense ark, in which the Nommos have re-entered a state of suspended animation necessary for interstellar voyages, might actually be the Saturnian moon Phoebe – an artificial body in a retrograde orbit around the ringed planet. Although some of Temple's detractors have ridiculed this fantastic notion, as was mentioned previously, there are those still attempting to establish contact with these higher intelligences in whatever alien environment they might currently be inhabiting. Put off by Leary and his telepathic Sirian experiments (performed during the "dog days" of summer, beginning on July 23), Temple has stated flat out that "there is nothing that [he] hates so much as drugs and drug-culture." I suppose this pretty much rules out any psychoactive substances associated with the shamanic experience and the idea of a chemical stargate whereupon one might access encyclopedic knowledge deposited in DNA. But what about another means of interacting with these beings? One person who was admittedly "staggered" by Temple's research was the late Robert Anton Wilson. In his mind-bending book *The Cosmic Trigger* (1976), the author goes to great lengths in exploring the Sirius phenomenon. In the course of his research he discovered that there is a deep connection between the western esoteric tradition and the Dog Star Sirius that Temple was either completely unaware of, or, for whatever reason, chose not to mention.

Scratching his head at this absence of Sirius-related arcana that might have bolstered his case, at one point an exasperated Wilson writes: "Over and over, one wishes that Temple had read a little in modern occultism to supplement his seven years of research..."

Is a God to Live in a Dog?

In reading the eroto-magical writings of Kenneth Grant (a former disciple of the infamous Aleister Crowley), Wilson was surprised to learn that the Sirius system was considered to be the source of the extraterrestrial current that powers Crowley's mystical Order of the Silver Star, the A.: A.: (*Argenteum Astrum*). The iconoclastic occultist Grant also claimed that the mysteries emanating from Sirius were the initiating forces of a branch of the O.T.O. (the quasi-Masonic *Ordo Templi Orientis*) that was specially founded to receive them. In developing this 'antenna' – such as occurs with a priestess in her oracular phase – adepts like Grant are said to be able to perceive faint impressions of astral level manifestations. Often these semi-tangible forms are loathsome in appearance – monstrous fancies resembling, perhaps, descriptions of the amphibious alien Nommos.

Another important element in the O.T.O. initiatory system is an entity known as Choronzon. As the guardian of the abyss, this "mighty devil" is regarded in occult cosmology as the final obstruction to initiation. However, in the dark doctrine, mastery over Choronzon also becomes the gateway to a "*mirroir fantastique*" code-named "the Nightside of Eden." Although Choronzon's speech is described as "weird and monstrous", it disperses the lost WORD, which cannot be heard unless one tunes in through the cosmic hiss. Such vibrations (channeled through the priestess) might be related to Temple's idea of harmonic resonance and of modulating the Anubis Field with one's thought patterns (though they could always try simply beating a drum, which is supposedly how the Dogon once communicated with the Nommos).

Aligned with the mysteries of "the Hidden God", and the occultist Aleister Crowley's tireless pursuit of admitting an extraterrestrial current into the human life-wave, is Timothy Leary's concept of eight potential neural circuits (or levels of consciousness). Leary anticipates that the eighth circuit will be meta-physiological (i.e. transcending biological existence) and, as such, its activation increases the possibility of contact with the "galactic Overmind." It might be a bit of a stretch to equate these eight circuits with assertions in a 'primitive' mythology that all humans are born with eight seeds stored in their shoulder-blades, but if there is a legitimate relationship between these complex Dogon concepts and our genetic inheritance (i.e. Nommo DNA), is it wise for occult 'gateways' like the Typhonian O.T.O. to engage in magical operations knowing that it could lead to the abhorrent things bestirring themselves?

From statements in the 1998 edition of his book, it's evident that Temple would prefer to distance himself from such "sinister" cults. Even so, as touched upon earlier, the truly ironic thing is that the author of *The Sirius Mystery* first became acquainted with the tribal lore of the Dogon in the mid-1960s when it was brought to his attention by his friend and mentor Arthur M. Young (founder of The Institute for the Study of Consciousness).

Temple further relates that his friend was introduced to the mystery by a fellow named Harry Smith, who gave him a Photostat translation of Griaule and Dieterland's *Le Renard Pâle*. Having been "struck by an irresistible urge" to investigate the enigma several years after first hearing about it, Temple received upon request a typescript of the anthropological report that was originally provided by Smith. Now, as it turns out, this Harry Smith was an avid experimenter with hallucinogenic drugs, as well as a card-carrying member of both Crowley's A.: A.: and the O.T.O. In fact, he even believed that he was the son of Aleister Crowley. Seriously.

Of course had Temple not been made aware of the French anthropologists' studies concerning an African tribe's inexplicable knowledge of Sirius B (and a putative third star) by his friend and tutor (who, as we now know, was himself turned onto the mystery by the Crowley devotee Harry Smith), perhaps eventually he might have heard about a colonial African-American farmer-astronomer named Benjamin Banneker. It seems that in the late 1700s Banneker used to talk a lot about his lucky star, Sirius, being a double star – something that he might have learned from his grandfather, Banneka, a first generation slave who, if not the stuff of legend, was once a member of the Dogon tribe.

Blair MacKenzie Blake has been studying, practicing, and writing about the western esoteric tradition for over twenty years. He is the author of two books: *Ijynx* and *The Wickedest Books in the World: Confessions of an Aleister Crowley Bibliophile*. In his 'other life', he is the writer/content manager for www.toolband.com and www.dannycarey.com.

Moai buried in sediment on Easter Island. (Photograph courtesy of R. Schoch and C. Ulissey.)

Moving the Moai

Easter Island as a possible psychokinetic laboratory

by *Robert M. Schoch, Ph. D.*

Easter Island is emblematic of ancient mysteries. Located in the South Pacific just south of the Tropic of Capricorn – some 3700 kilometers (2300 miles) west of mainland Chile and over 2000 kilometers (over 1200 miles) from any other inhabited island – it is often considered the most remote populated location on Earth. Also known as Rapa Nui, Isla de Pascua, and Te Pito Te Henua, the latter name has been translated as "the navel of the world," "navel and uterus," or "ends of the land," apparently referring to both the isolation of the island, and the volcanic craters that compose it. The island is triangular in shape, and so small that theoretically one can walk around its entire coast in less than two days. But, within this perimeter mysteries abound.

The island remained shrouded in secrecy, unknown to the Western world, until its discovery by Jacob Roggeveen on Easter Sunday, 1722. The giant stone sculpted heads and torsos, known as moai, that are found in the hundreds on the island are megalithic mysteries that boggle the imagination. The average height and weight of the moai has been estimated at 4 meters (13 feet) and about 12 to 13 tons, and the largest erected moai are up to 10 meters (33 feet) tall and are estimated to weigh up to 75 tons. An unfinished moai still in the quarry is some 21 meters (69 feet) long with an estimated weight of perhaps 250 tons. How and why were they carved? By whom and when? These are still unanswered questions; to this day the moai have eluded our best efforts to decipher their true meaning.

The standard modern "solution" to the enigma of Easter Island is that the earliest of the hundreds of moai were constructed after the first Polynesian settlers arrived on the island about 1500 years ago, and moai carving continued until around the seventeenth century, ceasing shortly prior to European contact. The moai, according to this scenario, were essentially part of an ancestor and burial cult. The production of moai ceased as the Easter Islanders deforested their land and consequently brought environmental devastation to their tiny island, ultimately sparking tribal warfare, an impoverished society, and even cannibalism. But does this story, which in recent decades has gained wide circulation in both scientific and popular circles, really stand up to scrutiny? A clear analysis of the evidence does not convincingly support the "natives destroyed their environment" scenario.[1] Rather, this may simply be a Western myth that both masks the destruction brought to the island and culture by European contact (which included notorious slave raids and effective genocide) and denies the true magnificence, antiquity, and high cultural and technological achievements of Easter Island civilization. It was with such thoughts in mind that in January 2010 my wife (Catherine Ulissey) and I, along with a few friends, undertook a short reconnaissance excursion to Easter Island to look into the mysteries for ourselves.

The Chronology and Dating of the Moai

The standard story of Easter Island does not detail the relative dating of the various moai, and in fact relegates all of these statues to a fairly late time period in accord with the hypothesis that the island was not inhabited before the first millennium A.D.

Studying the moai with a geological eye, I was particularly impressed by the varying degrees of weathering and erosion seen on different moai, which could be telltale signs of major discrepancies in their ages. The levels of sedimentation around certain moai also impressed me. Some moai have been buried in up to an estimated 6 meters (20 feet) of sediment, or more, such that even though they are standing erect, only their chins and heads are above the current ground level. Such high levels of sedimentation could occur quickly, for instance, if there were catastrophic landslides, mudflows, or possibly tsunamis washing over the island, but I could not find any such evidence (and landslides or tsunamis would tend to shift and knock over the tall statues). Rather, to my eye, the sedimentation around certain moai suggests a much more extreme antiquity than most conventional archaeologists and historians believe to be the case.

Not only does sedimentation around the statues suggest a longer and different chronology than conventionally accepted, but so too do weathering and erosion patterns, and stylistic considerations. Although on one level most of the moai are stylistically similar and even stereotypic, at another level each is unique and they could, I believe, be categorized according to stylistic considerations. The moai should, in addition, be sorted according to lithology (stone type) as well as weathering and erosion levels (taking orientation and relative exposure to the elements into account). Another key to solving the problem will be to compare weathering, erosion, and sedimentation rates in historical times. I have begun to gather photographs of various moai and landforms on Easter Island taken over the last 130 years so as to compare them geologically to their conditions today, and in this way attempt to get a quantitative handle on weathering, erosion, and sedimentation rates. So far, it seems that sedimentation over the last century has been on the whole relatively modest.

Even on our preliminary reconnaissance trip I found evidence that the earliest moai may have differed materially from the later moai. The earliest moai appear to have been more finely worked from hard basalts as compared to the volcanic tuffs from which most moai, dated to later periods, were carved. The few surviving basalt moai have been found at deeper stratigraphic levels below other moai and the platforms upon which later moai were erected, or basalt moai were reused in later structures – thus indicating the basalt moai are among the earliest on the island. Furthermore, at least one of the basalt moai (now housed in the museum on Easter Island) is of a very strange form; with an elongated head and well-defined breasts it is often considered a female while virtually all other moai are apparently males.

Another major puzzle, which is directly applicable to the chronology and dating of the moai, is the matter of where various moai were quarried. Quarries on the rim of the volcanic crater, where large moai were carved from the volcanic tuffs, are well exposed and still contain partially carved moai in place. I inspected these quarries carefully. However, the quarries where the few basalt moai were carved have never been definitively located, despite the small size of the island. Based on the geology of Easter Island, I expect that any suitable basalt deposits would occur lower in the stratigraphic section, so low in fact that they might currently be under sea level off the coast of the island. That is, the basalt quarries might be under water. How could this be? It is difficult to conceive that the ancient Easter Islanders were quarrying rock below sea level. Alternatively, the "lost basalt quarries" might be under sea level now because they are of extreme antiquity, and thus the basalt moai carved from them are extremely ancient. Sea levels have risen dramatically since the end of the last ice age, some ten thousand or more years ago, and if the basalt moai were quarried along the coast of Easter Island from areas

Female basalt moai now in the museum on Easter Island.
(Photograph courtesy of R. Schoch and C. Ulissey.)

since inundated by the sea, this could help to date the basalt moai, and is immediately suggestive that they are thousands of years older than conventionally believed to be the case.

The key to solving the puzzle of the basalt moai is to locate the quarries. If they are found under water along the coast, their depth and geological setting will shed light on their age, and could lead to a major revolution in our thinking about the antiquity, origin, importance, and sophistication of Easter Island culture. For one thing, while the later standard moai carved of volcanic tuffs could be relatively easily cut out of the rock using primitive tools found in abundance on the island, the same cannot be said for the apparently earlier and more sophisticated basalt moai, which may have formed the model and set the standard for the later volcanic tuffs moai.

A fascinating Easter Island legend relates that one of the founders, King Hotu-Matua, came from a land called Hiva that was said to be slowing sinking below the sea.[2] Is this suggestive of a people fleeing rising sea levels, perhaps thousands of years ago at the end of the last ice age when sea levels were rising worldwide? Was the initial colonization of Easter Island prior to the final, highest, post-ice age rises in sea level, such that when first colonized it was a larger and more prominent island? Such a scenario would be compatible with the early carving of basalt moai from quarries along the coast that have since been submerged.

The idea that the basalt moai may have been quarried from areas that are now under water is not based on mere speculation. While on the island we heard of reports that when Jacques-Yves Cousteau and his divers visited the island they came across some very rectangular looking holes or cavities in the basalt layers located off the coast. Could these indeed be ancient quarries? We began to track down leads as to divers, still on the island, who were with Cousteau's team and are willing and able to take us to these sites. I hope to dive and explore the possible quarries for myself on a return trip.

How Were the Moai Moved?

To this day, no one is absolutely certain how the moai were transported. Numerous theories involving brute manpower, ropes, wedges, levers, sledges, rollers, and so forth have been proposed. However, as Francis Mazière has perceptively pointed out,

> One of the great problems that dominate Easter Island archeologically [sic] is the question of how the statues were moved to the ahu [ceremonial platforms upon which they were erected], some of which are miles from the quarry. This problem has never been solved, and even the work of the Norwegian expedition of 1956 brought no answer, for [Thor] Heyerdahl's attempt at moving a moaï was in no way conclusive for . . . it was one of the smallest, and because it was dragged along by means of ropes over a very particular kind of ground that exists nowhere on the island except at Anakena – sandy ground without any rocks sticking out of it.[3]

Some traditional Easter Island legends state that the moai "walked" from the quarry to their final locations. This has been interpreted as indicating that the statues were raised upright, perhaps supported in slings hung from poles above, and then pivoted or rocked back and forth on their bases, thus moving them from one place to another – that is, they "walked." But was this really or always the case?

Another widespread Easter Island tradition is that the moai were transported by "mana," that is by supernatural forces or powers. As Katherine Routledge, who carried out extended exploration of Easter Island in 1914-1915, wrote:

> The natives are sometimes prepared to state that the statues were thrown down by human means, they never have any doubt that they were moved by supernatural power. We were once inspecting an ahu built on a natural eminence, one side was sheer cliff, the

Moai on an ahu. (Photograph courtesy of R. Schoch and C. Ulissey.)

> other was a slope of 29 feet [9 meters], as steep as a house roof, near the top a statue was lying. The most intelligent of our guides turned to me significantly. 'Do you mean to tell me,' he said, 'that that was not done by *mana*?'[4]

Alfred Métraux (1902-1963, a Swiss ethnologist who studied Easter Island extensively) wrote: "When the Easter Islanders of today are asked about the means by which the statues were transported, they only say: 'King Tiikoihu, the great magician, used to move them with the words of his mouth.'"[5]

The concept of "mana" is not generally considered scientific and thus is simply dismissed by most Western archaeologists and historians, but does it mask something we do not know? Perhaps a profound mystery that has yet to be uncovered? I take a serious scientific interest in the paranormal and, based on my research, I have concluded that there is strong evidence in some cases for the reality of psychokinesis (that is, the movement of matter via mental means alone; sometimes referred to as telekinesis).[6] Some of the best-documented cases of psychokinesis (once we have excluded fraud and delusion) are relatively spontaneous and unplanned, such as so-called poltergeist phenomena (in modern terminology often referred to as recurrent spontaneous psychokinesis, or RSPK).

Here I will explore the possibility that parapsychological insights into modern psychokinesis, combined with a particular theory of gravity, may provide a major clue to the mechanism underlying psychokinesis, and ultimately perhaps how the moai were moved. Admittedly this discussion is non-conventional, even radical and outrageous to some, but I believe it is at least heuristic to consider.

Gravity as an Entropic Force, and a Theory of Psychokinesis

In a recent paper Erik Verlinde (Institute for Theoretical Physics, University of Amsterdam) has developed the idea that gravity is not a fundamental force, but rather "an entropic force caused by changes in the information associated with the positions of material bodies".[7] Verlinde develops this theory in the context of a holographic concept of the universe, where the commonly understood structure of space-time is in fact an emergent property of information that can be visualized as stored on a surface or screen that wraps around the universe. In this conception mass emerges from information on the holographic screen surrounding a region (even if we conceptualize that region as the entire known universe). Note: Entropy is the degree of disorder within a system.

In the last paragraph I have tried to do justice, in quite abbreviated form, to Verlinde's thinking. Now I will develop it a bit further, and I will first state that my thoughts are an outgrowth of reading Verlinde's paper; but secondly, he may well not agree with my analyses, interpretations, and speculations!

To say that gravity is an entropic force is fundamentally saying that gravity as we observe it arises from the tendency of a system (which could be the entire universe) to statistically increase in entropy. Again, entropy is basically disorder within a system, and by the second law of thermodynamics, the entropy of an isolated non-equilibrium system will increase over time (that is, its order decays). A common everyday example of an entropic force is the pressure of a gas as the particles (atoms and molecules) spread out; a cellular example is the pressure of an osmotic gradient. If gravity is an entropic force, essentially the gravitational force is the tendency of objects (whatever objects are, as they may be epiphenomena) to move to positions of maximum disorder. Disorderly systems carry less information and orderly systems carry more information. Gravity can be equated with the tendency toward increasing entropy, which is increasing disorder, which in turn can be equated with minimizing informational content. Ultimately to "overcome" gravity, or act against the force of gravity, entails a change in the information content of a local system. In the everyday common "real world" we might change the information content of an object, say a stone, by physically lifting it up and placing it on a shelf. If it should be disturbed and fall down off the shelf, the local information content (ultimately encoded and stored on a holographic surface) would change once again.

In an ultimate sense, in a system completely lacking in information (no data encoded on the holographic screen), there would be no gravitational force as there could be no entropic force since minimal information and thus maximal disorder (entropy) will be the state of the system. One could visualize such a system as completely collapsed down to a single dimensionless point, but since there is no information in the system, there would be no mass (since mass emerges from information) or space-time. If information were to be inserted into the system, mass and space-time would emerge, and also gravity as an entropic force countering the information (statistically driving the system back to maximum entropy and minimum information). Information would create the universe as we know it, and inserting increasing amounts of information into the system would cause the universe to expand outward even as gravity countered the information tendency.

The most fundamental aspect of the universe, by this conception, is information. Information may be equated with thought or mind or mental constructs independent of any material everyday conception (certain aspects of parapyschological studies strongly support the notion of the independence of thought or mind from any kind of material object, like a physical brain). The universe of mass and energy and forces, as we experience it on an everyday practical level, may have had its origin in a thought that inserted information into an otherwise blank (data free) holographic system. (I will leave the question unanswered as to where the initial thought came from, although I cannot help but think of the famous lines from Christian scriptures, "In the beginning was the Word [thought, information]: the Word was with God and the Word was God." John 1:1.) Inserting more thoughts, more information, expands and changes the system, allowing it to develop or evolve. Furthermore, thoughts at a certain level are self-perpetuating (here memory of thoughts, encoding and preserving of information, may play a key role) and self-expanding as they build, grow, and elaborate upon themselves, and in this scenario all the while building and developing the "material world" and universe, as we know it. Thinking entities who accept the epiphenomena of the everyday world as they experience it on a mundane level will put out more thoughts (habits of thinking) which indeed reinforce not just their status quo conception of the world but also literally their experiential world. Once thinking entities accept a particular worldview (including a conception of the world at the most fundamental level, such as the basic laws of physics) it will be self-perpetuating until or unless different information (thoughts) is inserted into the system at the most fundamental level.

So, by this hypothesis, inserting or changing information by mental means, by thoughts alone, can change (add to, rearrange, and integrate) the data on the holographic screen that "creates" the universe as we know it and thus we can change the universe, or some localized region or sector of the universe. In particular, changing information may allow gravity to be overcome, and this is precisely what we see in genuine cases of psychokinesis. However, the key (or "trick") is actually changing the information. It seems clear that it is not enough (at least for the vast majority!) to simply "think" on a conscious level about producing a psychokinetic event, but a very deep unconscious or subliminal "belief" (for lack of a better term at the moment) must be present for information to be changed such that manifestations occur in the everyday "real" world. Furthermore, successfully injecting or modifying information, so as to manifest observable changes, may also require a collective level of belief, or at the least overcoming the collective disbelief (perhaps by the force of the "belief" of a single or few individuals), in a local system.

Among the best-documented cases of mind over matter (the movement of objects by non-physical means) are so-called poltergeist instances. In many such cases it appears that objects are moved paranormally (items thrown through the air or falling off of shelves when nobody is close enough to reach them, and no physical means are apparent to account for the movement of the objects) due to the unconscious emanations of an emotionally or psychologically disturbed individual. In these instances the person who is the source or focus of the poltergeist manifestations may have, at a deep unconscious level, broken ties with everyday reality and as a result he or she is able to successfully change information in the holographic system, which results in manifestations in the "real world" that defy normal gravitational constraints. Other cases of genuine levitation, movements of objects, and other gravity-defying and psychokinetic phenomena (even to move an object horizontally, without perceptively lifting it vertically, can involve overcoming the force of gravity to some extent), have been documented in séance types of settings where a critical mass of the participants (it is not clear if 100% are always necessary) have apparently either allowed themselves at a deep level to "believe" that such things can occur, or at least gotten to an open attitude of suspension of disbelief (so as not to unduly hinder the process). As a case in point along these lines, in 1869-1870 a committee of the London Dialectical Society investigated so-called spiritualistic phenomena, including experimenting with séances. Some of the investigators were apparently skeptical, but open-minded, and on numerous occasions a heavy table, for instance, was witnessed to move in various directions without any human contact or other mechanical or physical explanation. Many other cases of levitations, movements of objects, and other physical phenomena are recorded in the serious literature.[8]

Another context where psychokinesis appears to be well documented is on the micro- or atomic level, for instance influencing the rolling of dice or the actions of random event generators.[9] In such cases, I hypothesize, it may be that it is easier to "believe" that psychokinesis can occur on a microscopic or atomic level, and therefore it is easier to insert or change information in the system, overcoming entropy, and effect real changes manifested in the observable world.

In some cases it has been claimed that individuals can manifest macroscopic psychokinetic effects by conscious will or volition. An example of such an individual was the Russian Nina S. Kulagina. I have to admit that I remain highly skeptical of such reports, and I cannot vouch for the authenticity of Kulagina's phenomena, but it is instructive to consider her case. Typical for Kulagina was the movement of small and light objects, such as earrings, rings, envelopes, etc., across a tabletop; a 1970 report stated that the largest object moved was a glass vase weighing 380 grams. In order to effect positive results, Kulagina had to be in a suitable psychological state. Furthermore,

> It was discovered that even the psychological state of the persons witnessing the experiment was of importance too. Interest, open mutual trust and friendly feelings are very helpful for the experiments. In cases when the initiative for the experiments did not come from the subject, when the circumstances were unfamiliar and the experiment organised in a different way, e.g., new apparatus, new objects, strangers present, then it is difficult to get good results. In such cases, the subject has to exert more energy, takes longer to prepare the test, and her strain during the experiment is greater... The circumstances grow more complicated if some of the people present do not believe in the success of the experiment or are sceptical of the results. Sometimes, although conditions seem favourable and the subject tried hard to carry out the experiment, the necessary psychological state could not be produced in the subject. In some cases the subject could identify those present who were hindering this by concealed doubts and scepticism. Provided that they then left the room, an atmosphere of confidence could be created and the experiment completed successfully. If after this initial success, witnessed by all present, the sceptics were readmitted, their presence no longer impeded results.[10]

The correct psychological state of not only Kulagina, but also everyone present, was very important to the success of the psychokinetic endeavors. According to the theory I am developing here, it is essential to adequately diminish or eliminate the individual and collective skepticism about whether such things as psychokinesis are possible before information can be adequately manipulated to effect a "real world" psychokinetic event. Once success has been achieved in certain contexts, the "beliefs" of those who observed the success might be enough to overcome the "doubts" of the skeptics. The limiting of successful psychokinesis to relatively light objects in the case of Kulagina may be both a function of what it was "believed" could be accomplished (it is perhaps easier to believe, at the deepest levels of the unconscious, that smaller and lighter objects can be affected) and the issue that it perhaps requires a higher degree of information manipulation to psychokinetically move a heavier object (weight being a measure of the entropy that needs to be overcome).

As a side note to the concept of gravity as an entropic force in a holographic system, essentially gravity would be equivalent to information decay. To maintain records in the form of data encoded on the holographic screen the data would have to be continually reinforced (perhaps by thought). But not only could it be reinforced, it could also be changed (either purposefully or inadvertently), and anyone accessing the record may receive varying degrees of veridical information depending on the context (in terms of an equivalence principle, if information can be added, then it can also be accessed and/or extracted or removed). This may explain aspects of psychical or mental paranormal phenomena (such as telepathy, retrocognition, and precognition) and also suggests that both the past and the future from a human perspective may be malleable as the information encoded on the holographic screen can be manipulated. For thought producing and receiving entities like us, the most easily accessible and manipulated information is that of the immediate local environment (local from our perspective; that is, close in time and space from an everyday perspective).

Easter Island as a Psychokinesis Laboratory

Returning to the moai of Easter Island, is there any truth to the legends of mana? Were the moai moved paranormally, by a form of psychokinesis, at least in part? Observing the moai on the island, especially around the major quarry, I could not help but get the impression that they had, at least in some cases, been popped out or tossed out of their resting places, as if by some semi-chaotic force. It is almost as if parts of the island had been subject to a poltergeist style phenomenon, with moai strewn about. Perhaps, and of course this is wild speculation, once a moai was partially carved (more or less roughed out), it was literally moved or thrown out of its quarry spot paranormally. The final positioning and raising of the moai on an ahu may have been accomplished by more normal means.

Easter Island was literally a world unto itself, isolated in space by thousands of kilometers and isolated in time by many centuries and millennia. Easter Island may have formed the perfect psychokinesis laboratory. The inhabitants developed their own collective psyche and worldview, one that was perfectly "believable" to them and included the ability to manipulate data on the local holographic screen and thus accomplish psychokinetic feats on moai weighing tens of tons. The Easter Islanders may have had 100%, or nearly 100%, investment in the belief system of the island. We can think of the entire island being involved in a giant séance where everyone is a true believer. The resulting phenomena may have been well beyond the possibilities that most people can even entertain in our Western twenty-first century culture.

So, were the moai moved paranormally? Did mana indeed have a role to play? I cannot say for certain, but I believe we should be open to the possibility. Oddly, the psychotherapist Werner Wolff (1904-1957) seriously suggested that the moai had flown through the air, but not via parapsychological mana, but by "natural mana"—that is by the explosions of the volcano where the moai were carved:

> Since great blocks of lava have been discharged by the volcanoes, why could such blocks not have been ejected after being shaped by sculptors in the workshop of the crater? This theory, in a way, would

Moai on an ahu. (Photograph courtesy of R. Schoch and C. Ulissey.)

> explain the so-called magic transportation, although imagining giant-statues flying through the air like ghosts seems to resemble a fantastic dream. Unconsciously, perhaps, each investigator has dismissed such a possibility; it does seem strange though that this angle has never been discussed. . . .
>
> But the idea of flying statues did not seem strange to the natives—the concept was within the sphere of their imagination. C. E. Fox reports such a concept from San Christobal in the Solomons: 'Levitation is described as common both in the case of sacred stones and of priests, some of whom were levitated large distances through the air.'[11]

As a geologist, I question whether moai were moved by volcanic power. Although carved of solid rock, I doubt that any moai

Unfinished moai in the quarry. (Photograph courtesy of R. Schoch and C. Ulissey.)

would have survived intact after being shot like a missile from the volcano and then crash-landing on the ground. And I wonder how many human sculptors might have been killed, or how any could have survived, working in a volcanic crater that was on the verge of erupting. And think of all the moai that may have shot out of the volcano only to land in the ocean surrounding the tiny island.

Interestingly, despite the emphasis on mana as the way the moai were moved, as recorded by Katherine Routledge, J. Macmillan Brown reported that the Easter Islanders did not even have a native word for mana. Brown resided on the island for five months in the early 1920s (some of the photos reproduced in his 1924 book are dated to 1922), and interviewed and worked with some of the same informants as did Routledge. Brown wrote, An...

> ...important *lacuna* [in the Easter Island vocabulary] is the word *mana*, which in Polynesian covered all that occult power which resides, not only in the supernatural and the ancestors, but in great living men and their deeds and their weapons and instruments. It was probably because there was no sense of this *aura* of the great and indeed no great individuality to embody it or ray it forth that the word dropped out of their Polynesian vocabulary.[12]

Thus, Brown is not contending that the Easter Islanders of his time did not believe in mana, but simply that they did not have, or no longer had, a native word for the concept. Indeed, one might speculate, this was because the mana was no longer manifested on Easter Island after the population had been decimated, and the indigenous culture all but eradicated, in the wake of European contact beginning in 1722.

Legends of Giants

Sometimes it is suggested that the moai were huge simply because the peoples inhabiting an ancient Easter Island were huge. Indeed there is some, if at this point only very scanty, evidence that besides the giant moai, there may have been other giants on Easter Island. Literal living giants, that is. C. F. Behrens, who accompanied J. Roggeveen on the voyage that marks the generally accepted European discovery of Easter Island in 1722, reported that...

> ...with truth, I might say that these savages are all of more than gigantic size. The men are tall and broad in proportion, averaging 12 feet [3.65 meters] in height. Surprising as it may appear, the tallest men on board of our ship could pass between the legs of these children of Goliath without bending the head. The women can not compare in stature with the men, as they are commonly not above 10 feet high [3 meters].[13]

What do we make of such a report? Could this be real? Interestingly, traditional legends from the island of Flores, Indonesia, describe the existence of a small race of diminutive people known as the Ebu Gogo.[14] Recently a new species of humans, *Homo floresiensis* – popularly referred to as "hobbits" – has been found on Flores that may fit the description of the Ebu Gogo.[15] *H. floresiensis* is based on sub-fossil remains from the end of the last ice age, as recent as 12,000 years ago. However, some serious researchers have hypothesized that hobbits survived as a relict population until possibly as late as the nineteenth century.[16] Likewise, could the early reports from Easter Island be factual? Stories of "giants" are reported in sacred texts, myths, and legends around the world. Might these legends have a basis in fact? Might a relict population of "giants" have survived on Easter Island up until European contact?

The discovery of solid physical evidence for "giants" on Easter Island would revolutionize our thinking about the ancient past, shedding new light on many myths and legends, and force us to rethink the development of humans through time.

Much to Learn

Exploring Easter Island firsthand, I quickly became convinced that the standard, and rather mundane, explanations for the island's remarkable indigenous civilization and their achievements do not necessarily hold up under detailed examination. Pursuing further research concerning the island and its "mysteries" (an admittedly overused term, but one that seems genuinely applicable here), we have much to learn from this small isolated spot in the Pacific. Easter Island and its world prior to European contact may have been very different, at a deep level, from that which most people today perceive as reality. Indeed, so different that it is extremely difficult, if not virtually impossible, to fully grasp. However in trying to comprehend the lost world of Easter Island we open up new possibilities and begin to expose the assumptions, beliefs, and biases that prevent us from truly understanding the ultimate nature of reality. I can think of nothing that is more profound and worthwhile.

Robert M. Schoch received a Ph.D. in Geology and Geophysics from Yale University, and since 1984 has been a full-time faculty member at the College of General Studies of Boston University. His books include the trilogy with R. A. McNally: *Voices of the Rocks, Voyages of the Pyramid Builders,* and *Pyramid Quest.* His most recent book is *The Parapsychology Revolution: A Concise Anthology of Paranormal and Psychical Research* (Compilation and Commentary by Robert M. Schoch and Logan Yonavjak, Tarcher/Penguin, 2008). Dr. Schoch's personal website is located at: www.robertschoch.com.

THE MILL RACE MONSTER

AN EXCURSION INTO THE OBSCURE ANNALS OF CRYPTOZOOLOGY

by *Neil Arnold*

During the early months of 2010 I was contacted by a man named Chris McDaniel. Chris lives in Columbus, Indiana, USA, and is an Animal Control Officer. In the past Chris worked for four years at an Indianapolis zoo, and is qualified in wildlife management. He also has a passion for cryptozoology – the study of hidden animals. In his email Chris asked me if I'd ever heard the tale of the Mill Race Monster. I originally thought he was speaking of a creature reported in the *Milwaukee Journal* of May 26th 1983, under the heading: "Chicago Angler Lands Millrace Monster". The story read as follows:

> The 'Millrace Monster', a mysterious canal dweller blamed for devouring ducks, has been caught with a hook and line. Some

> residents thought they had their own Loch Ness monster after two people reported seeing ducks snatched from below the surface of the Millrace, a canal off the Kankakee River where a mill wheel once turned.
>
> Arnold Chipusa of the Illinois Department of Conservation said the so-called 'monster' could be a snapping turtle, but more likely was a huge northern pike. And then angler Jim Pecoraro of Chicago landed such a fish, a 41-inch specimen weighing 20 pounds, 2 ounces. The largest pike ever caught in the state was 22 pounds, 12 ounces.

The following year, on February 21st 1984, *The Argus Press* ran a story titled "Monster Fish' Makes Life Rough For Migrating Fowl":

> A monster fish with a taste for duck is making life rough for migrating fowl seeking a place to rest on the Kankakee River near here, fishermen say.
>
> Several fishermen told bait shop owner Howard Curtis that a large northern pike was "taking swipes" at ducks on the water. "This fellow, a good customer of mine, came in yesterday and he says, 'Howard, I couldn't tell what kind of fish he was, but he was really after them ducks', said Curtis.
>
> ... Many fishermen think the big fish might be a cousin of an upstream northern, dubbed the 'Mill Race Monster', which lurked in Kankakee River holes near Wilmington last spring. That fish was credited with plucking several ducks off the surface of the Mill Race pond last April.

I emailed the story I'd written up to Chris, and although fascinated by it, he replied, "It's interesting to note that there is 'another' Mill

Race Monster, but the one of which I speak is of greater interest. It also happens to be so obscure that no-one on the internet, or in books, appears to have picked up on it, however, I was fortunate enough to investigate the case and am now passing all the information on to you."

A Strange and Terrifying Creature...

The information I received was far more intriguing than the Kankakee canal dweller I'd found. And so the story goes...that, for just over a fortnight in November of 1974, at a wooded area known as Mill Race Park, on the edge of Columbus, Indiana, a strange and terrifying creature was said to have spooked several locals and attacked a number of vehicles, leaving huge scratch marks on the paint work. The beast became known as the Mill Race Monster and it was described as being a hair-covered bipedal creature with a pale green face. Mill Race Park was once known as Death Valley due to its population of rats and disease. Additionally, houses in the area were prone to flooding, making it all-in-all a rather inhospitable place to reside. However, during the 1960s the area was purchased by the city and underwent a positive transformation. Mill Race covers more than eighty acres and its covered bridge harkens back to the days when such constructions were covered in order for the planks of wood to be protected.

The first report to surface of a strange presence in the park was featured in a local newspaper on 2nd November under the heading, "Women report seeing 'beast' in Mill Race":

> They swear it's true. Six young women told city police Friday in two separate reports that they had seen a "Mill Race Monster" – once in broad daylight. And whatever it is, the six are agreed it is: 1) Green 2) Hairy 3) Large, about six-feet tall and walking upright, and 4) Has claws.

Four of the women, whose names police did not list, said they saw "the thing" about 3pm Friday, near the paved boat ramp in the park along the White river at the west end of Fifth street. The other two said it jumped on the hood of their car, leaving scratch marks in the paint. This was about 11:45 pm. No-one was hurt, only frightened, police said.

Officers searched the area on both occasions but found nothing. They are hoping it's only a misguided hobgoblin who lost its sense of timing and doesn't know Halloween is over for another year. But, just the same, they warned other residents to take care and keep an eye out for the "monster".

The women described the creature they encountered near the boat dock as "the weirdest thing", saying the "big, black blurry-looking thing came towards the car", noting that its odd face was definitely

Monster

Women Report Seeing 'Beast' in Mill Race

They swear it's true. Six young women told city police Friday in two separate reports that they had seen a "Mill Race Monster" — once in broad daylight.

And whatever it is, the six are agreed, it is: (1) Green, (2) Hairy, (3) Large, about six feet tall and walking upright, and (4) Has claws.

Four of the women, whose names police did not list, said they saw "the thing" about 3 p.m. Friday near the paved boat ramp in the park along White river at the west end of Fifth street. The other two said it jumped on the hood of their car, leaving scratch marks in the paint. That was about 11:45 p.m. No one was hurt, only frightened, police said.

Officers searched the area on both occasions but found nothing. They are hoping it's only a misguided hobgoblin who lost its sense of timing and doesn't know Halloween is over for another year. But, just the same, they warned other residents to take care and keep an eye out for the "monster."

not a mask. “It did not look anything like a mask or a person playing a prank. The face was pale-looking…greenish-white”.

Four days later, on November 6th, the monster was back in the news. Under the title “Mill Race Monster Makes Appearance After Voting?”, the brief report noted that…

> …The Mill Race Monster has returned, apparently stepping out of the shadows Tuesday long enough to scare two park-goers and possibly cast his vote in local elections.
>
> County police reported two men in their 20s saw a large, hairy “thing” hiding behind trees near the covered bridge about 4 pm and again shortly before midnight. The men, who did not give police their names, said after seeing the monster Tuesday afternoon, they returned with binoculars that evening for a closer look. They told police the monster was found, and eventually chased them from the park.
>
> One of the men suggested an explanation for the newly-found monster, that it had been left here in the early fall by visitors from another planet. At that time, Columbus, and much of the Midwest was flooded by sightings of Unidentified Flying Objects (UFOs).
>
> One political know-it-all when hearing about the monster sighting Tuesday night, suggested it was a publicity-seeking candidate trying to begin an early campaign for office in 1976.

By the 8th November public interest in the monster had blossomed, and several monster hunters had invaded the wooded park in hope of catching a glimpse of the green-faced phantom. The local newspaper headline on that day read “Mill Race Monster Hunt Out Of Hand”, after city park director Robert Gilligan had issued warnings to anyone wishing to set up vigil in the park in hope of capturing the elusive creature. Local authorities had reported

Tonight: Cooler
Tomorrow: Partly Sunny
— Price 15 Cents

THE REPUBLIC

No Week End Monster Sightings

Today's situation report on the Mill Race park Green Monster shows:

— Mill Race park grounds are to remain barricaded and closed from dusk until dawn to protect the monster's public.

queries from out-of-town reporters to provide details on the "monstrous situation."

Robert Gillikin, parks director, said today Mill Race park will remain closed at night as a safety precaution for the curiosity-seekers wishing to see the monster, and for the monster itself.

The park was barricaded and closed beginning Friday night.

Gillikin said youths were carrying knives and clubs and said they might hurt each other or the monster if it is caught.

He did not know when the park would reopen.

Gillikin said park officials at the barricade did not see the monster.

The police department received no reports saying the monster was sighted anywhere during the week end.

Transcript Changed In Watergate Trial

WASHINGTON (UPI) — used to finance the operation. Committee. Thus it could

that several individuals had set up camp in the park the night after the November 6 story broke. Police had discovered several people sitting in vehicles, whilst other persons were encountered in the woods. The biggest concern for the police was the fact that a majority of the individuals staking the area out were armed with knives and other weapons, but the authorities had begun to believe that the 'monster' was in fact a prankster and they didn't want the jester to be apprehended by local hunters in case he was battered, or hacked to death by those thinking he was the beast. Robert Gilligan decided that the park would have to be closed of a night, after more than twenty-five vehicles were found parked in the area at 11pm on the Thursday.

According to the news report, a dog-catcher, Rick Duckworth, had been the latest witness to the beast. Whilst on patrol on the 8th November, Rick stated he'd seen what appeared to be a person "wrapped in blankets and wearing a mask". The details certainly lacked the monstrous drama of the original reports. The following day interest in the Mill Race hysteria seemed to be waning. "Green Monster No Danger, People Are", read the headline, the report commenting on the disappointing no-show of the phantom humanoid: "The Green Monster of Mill Race park did not show up

Friday night. City police received a call on the monster's whereabouts but it was a false alarm."

This report however, didn't seem to tie in with the details given in previous reports: not only did it call the park director "Mr Gillikin" but it also stated that more than "100 cars" had been counted in the park, which led to the director closing the park for several nights.

At the time *The Indianapolis Star* was said to have devoted three columns on page one of its Saturday issue to the monster craze.

The Republic of Monday, November 11th reported "No Week End Monster Sightings" after the park had been barricaded shut. Columbus Police Dept. had established a "monster control officer", named Kenneth St. John, to deal with any queries from journalists from newspapers countrywide interested in the story. Calls had been taken from Chicago, Indianapolis and Louisville in connection with what the press and police were now calling a "monstrous situation".

Although a small snippet appeared in the local newspaper on the 12th, just to reiterate the park closure story, it appears that no more sightings of the Mill Race wraith were reported. And so, the monster faded into obscure folklore until Chris McDaniel investigated the sightings. McDaniel spoke to retired Sgt. Dennis Knulf, who stated quite categorically that the Mill Race panic was simply a hoax. Chris questioned this however, unsure as to how a hoaxer could leap onto vehicles, not only avoiding injury to himself, but also leaving huge scratch marks in the paint work.

HOLIDAY HOAX?
Former Columbus policeman recounts 1974 Halloween sightings of Mill Race Monster, Page A5
MONSTER PARTY!
Loch Ness monster, Sasquatch seen living it up at Vegas casino
OPRAH ABDUCTED BY ALIENS!

On October 31st 2003, some twenty-nine years after the brief scare, *The Republic* ran a short article titled "Residents laugh about '74 monster hoax", with Sgt. Knulf commenting "We joked about it…we think it was just someone playing a hoax, but we don't know for sure." However, Knulf was also quoted as saying that after the attack on the vehicle, strange footprints were discovered in the mud at the park. Knulf, despite his scepticism, also commented that, "Some people were pretty scared after seeing the thing..." But when asked if he'd ever seen the creature himself, he paused for a while, and replied, "I ain't sure...I talked to some people who were there who said they saw it...but I don't talk about it."

Whatever lurked around Mill Race Park for that short time scared some families enough to refuse to visit the area with their children for a picnic. The scare also attracted visitors from as far afield as Chicago, who obviously felt that the four-hour drive was worth a part in the monster-mania.

Bigfoot By Any Other Name?

Chris McDaniel also looked into the possibility that the humanoid was a Bigfoot. However, the Bigfoot Research Organisation (BFRO), despite methodical research into the phenomenon of hairy bipeds resulting in a list of more than sixty reports throughout Indiana dating back to the 1970s, mentions only three sightings at Bartholomew County. The earliest of the trio only dates back to the winter of 1992 when two men, travelling on the 'Youth Camp Road' around the Bartholomew/Brown county line west of Columbus at night, picked up a white-furred, bipedal creature in the headlights as it emerged from woodland between fifty and eighty yards ahead. It crossed from the left and slipped into the woods on the other side of the road.

In April 2001 near the State Road 46, an eighteen-year-old male was studying in his room at 1:30am and went outside to his vehicle to get a pencil. When stepping outside, a tremendous cry emanated from the park directly opposite the house. The youth observed a tall, man-like creature running through the thicket. Five years later on 7th June 2006, around ten miles from Columbus, a female witness walking her puppy at 2:30am heard a peculiar, eerie howl which sounded like no known native animal. There was also a sound as if something large was crashing through the trees.

So, the mystery of the Mill Race Monster never fully reared its ugly green face. What or who this critter was we'll never know, but one thing is for sure: this extremely obscure spook story can now be dragged from the shadows of folklore, unlike the man or thing behind the mask.

Neil Arnold was born in Kent, England in 1974 and has, since the age of 9-years-old, been interested in monsters after seeing the movie *The Legend of Boggy Creek*. He is the author of *Monster! The A-Z of Zooform Phenomena* (http://zooform.blogspot.com), which Loren Coleman voted the 2007 Fortean Zoology Book of the Year, *Mystery Animals of Kent, Paranormal London* and *Paranormal Kent*. He writes for magazines, has appeared on television and radio, is a full-time monster-hunter and runs Kent Big Cat Research (http://kentbigcats.blogspot.com).

"Cthulhu and R'lyeh." (Dominique Signoret)

Calling Cthulhu

H.P. Lovecraft's Magick Realism

by *Erik Davis*

Consumed by cancer in 1937 at the age of 46, the last scion of a faded aristocratic New England family, the horror writer Howard Phillips Lovecraft left one of America's most curious literary legacies. The bulk of his short stories appeared in *Weird Tales*, a pulp magazine devoted to the supernatural. But within these modest confines, Lovecraft brought dark fantasy screaming into the 20th century, taking the genre, almost literally, into a new dimension.

Nowhere is this more evident than in the loosely linked cycle of stories known as the Cthulhu Mythos. Named for a tentacled alien monster who waits dreaming beneath the sea in the sunken city of R'lyeh, the Mythos encompasses the cosmic career of a variety of gruesome extraterrestrial entities that include Yog-Sothoth,

Nyarlathotep, and the blind idiot god Azathoth, who sprawls at the center of Ultimate Chaos, "encircled by his flopping horde of mindless and amorphous dancers, and lulled by the thin monotonous piping of a demonic flute held in nameless paws."[1] Lurking on the margins of our space-time continuum, this merry crew of Outer Gods and Great Old Ones are now attempting to invade our world through science and dream and horrid rites.

As a marginally popular writer working in the literary equivalent of the gutter, Lovecraft received no serious attention during his lifetime. But while most 1930s pulp fiction is nearly unreadable today, Lovecraft continues to attract attention. In France and Japan, his tales of cosmic fungi, degenerate cults and seriously bad dreams are recognized as works of bent genius, and the celebrated French philosophers Gilles Deleuze and Felix Guattari praise his radical embrace of multiplicity in their magnum opus *A Thousand Plateaus.*[2] On Anglo-American turf, Lovecraft has been enshrined in the august Library of America, while a passionate cabal of critics fill journals like *Lovecraft Studies* and *Crypt of Cthulhu* with their almost talmudic research. Meanwhile both hacks and gifted disciples continue to craft stories that elaborate the Cthulhu Mythos. There's even a Lovecraft convention – the NecronomiCon, named for the most famous of his forbidden grimoires. Like the gnostic science fiction writer Philip K. Dick, H.P. Lovecraft is the epitome of a cult author.

The word "fan" comes from *fanaticus*, an ancient term for a temple devotee, and Lovecraft fans exhibit the unflagging devotion, fetishism and sectarian debates that have characterized popular religious cults throughout the ages. But Lovecraft's "cult" status has a curiously literal dimension. Many magicians and occultists have taken up his Mythos as source material for their practice. Drawn from the darker regions of the esoteric counterculture – Thelema and Satanism and Chaos magic – these Lovecraftian mages actively seek to generate the terrifying and atavistic encounters that Lovecraft's protagonists stumble into compulsively, blindly or against their will.

Secondary occult sources for Lovecraftian magic include a handful of different "fake" editions of the *Necronomicon*, a few rites included in Anton LaVey's *The Satanic Rituals*, and a number of works by the loopy British Thelemite Kenneth Grant. Besides Grant's Typhonian O.T.O. and the Temple of Set's Order of the Trapezoid, magical sects that tap the Cthulhu current have included the Esoteric Order of Dagon, the Bate Cabal, Michael Bertiaux's Lovecraftian Coven, and a Starry Wisdom group in Florida, named after the nineteenth-century sect featured in Lovecraft's "Haunter of the Dark." Solo chaos mages fill out the ranks, cobbling together Lovecraftian arcana on the Internet or freely sampling the Mythos in their chthonic, open-ended (anti-) workings.

This phenomenon is made all the more intriguing by the fact that Lovecraft himself was a "mechanistic materialist" philosophically opposed to spirituality and magic of any kind. Accounting for this discrepancy is only one of many curious problems raised by the apparent power of Lovecraftian magic. Why and how do these pulp visions "work"? What constitutes the "authentic" occult? How does magic relate to the tension between fact and fable? As I hope to show, Lovecraftian magic is not a pop hallucination but an imaginative and coherent "reading" set in motion by the dynamics of Lovecraft's own texts, a set of thematic, stylistic, and intertextual strategies that constitute what I call Lovecraft's Magick Realism.

Magical realism already denotes a strain of Latin American fiction – exemplified by Borges, Gabriel Garcia Marquez, and Isabel Allende – in which a fantastic dreamlike logic melds seamlessly and delightfully with the rhythms of the everyday. Lovecraft's Magick Realism is far more dark and convulsive, as ancient and amoral forces violently puncture the realistic surface of his tales. Lovecraft constructs and then collapses a number of intense polarities – between realism and fantasy, book and dream, reason and its chaotic Other. By playing out these tensions in his writing, Lovecraft also reflects the transformations that darkside occultism has undergone as it confronts modernity in such forms as psychology, quantum physics, and the existential groundlessness of being. And by embedding all this in an intertextual Mythos of profound depth, he draws the reader into the chaos that lies "between the worlds" of magick and reality.

A Pulp Poe

Written mostly in the 1920s and '30s, Lovecraft's work builds a somewhat rickety bridge between the florid decadence of *fin de siècle* fantasy and the more "rational" demands of the new century's science fiction. His early writing is gaudy Gothic pastiche, but in his mature Chtulhu tales, Lovecraft adopts a pseudodocumentary style that utilizes the language of journalism, scholarship, and science to construct a realistic and measured prose voice which then explodes into feverish, adjectival horror. Some find Lovecraft's intensity atrocious – not everyone can enjoy a writer capable of comparing a strange light to "a glutted swarm of corpse-fed fireflies dancing hellish sarabands over an accursed marsh."[3]

But in terms of horror, Lovecraft delivers. His protagonist is usually a reclusive bookish type, a scholar or artist who is or is known to the first-person narrator. Stumbling onto odd coincidences or beset with strange dreams, his intellectual curiosity drives him to pore through forbidden books or local folklore, his empirical turn of mind blinding him to the nightmarish scenario that the reader can see slowly building up around him. When the Mythos finally breaks through, it often shatters him, even though the invasion is generally more cognitive than physical.

By endlessly playing out a shared collection of images and tropes, genres like weird fiction also generate a collective resonance that can seem both "archetypal" and cliched. Though Lovecraft broke with classic fantasy, he gave his Mythos density and depth by building a shared world to house his disparate tales. The Mythos stories all share a liminal map that weaves fictional places like Arkham, Dunwich, and Miskatonic University into the New England landscape; they also refer to a common body of entities and forbidden books. A relatively common feature in fantasy fiction, these metafictional techniques create the sense that Lovecraft's Mythos lies beyond each individual tale, hovering in a dimension halfway between fantasy and the real.

Lovecraft did not just tell tales – he built a world. It's no accident that one of the more successful role-playing games to follow in the heels of *Dungeons & Dragons* takes place in "Lovecraft Country." Most role-playing adventure games build their worlds inside highly codified "mythic" spaces of the collective imagination (heroic fantasy, cyberpunk, vampire Paris, Arthur's Britain). The game *Call of Cthulhu* takes place in Lovecraft's 1920s America, where players become "investigators" who track down dark rumors or heinous occult crimes that gradually open up the reality of the monsters. *Call of Cthulhu* is an unusually dark game; the best investigators can do is to retain sanity and stave off the monsters' eventual apocalyptic triumph. In many ways *Call of Cthulhu* "works" because of the considerable density of Lovecraft's original Mythos, a density which the game itself also contributes to.

Lovecraft himself "collectivized" and deepened his Mythos by encouraging his friends to write stories that take place within it. Writers like Clark Ashton Smith, Robert Howard, and a young Robert Bloch complied. After Lovecraft's death, August Derleth carried on this tradition with great devotion, and today, dozens continue to write Lovecraftian tales.

With some notable exceptions, most of these writers mangle the Myth, often by detailing horrors the master wisely left shrouded in ambiguous gloom.[4] The exact delineations of Lovecraft's cosmic cast and timeline remain murky even after a great deal of close-reading and cross-referencing. But in the hands of the Catholic Derleth, the extraterrestrial Great Old Ones become elemental demons defeated by the "good" Elder Gods. Forcing Lovecraft's cosmic and fundamentally amoral pantheon into a traditional religious framework, Derleth committed an error at once imaginative and interpretive. For despite the diabolical aura of his creatures, Lovecraft generates much of his power by stepping beyond good and evil.

The Horror of Reason

For the most part Lovecraft abandoned the supernatural and religious underpinnings of the classic supernatural tale, turning instead towards science to provide frameworks for horror. Calling Lovecraft the "Copernicus of the horror tale," the fantasy writer Fritz Leiber Jr. wrote that Lovecraft was the first fantasist who "firmly attached the emotion of spectral dread to such concepts as outer space, the rim of the cosmos, alien beings, unsuspected dimensions, and the conceivable universes lying outside our own spacetime continuum."[5] As Lovecraft himself put it in a letter, "The time has come when the normal revolt against time, space, and matter must assume a form not overtly incompatible with what is known of reality – when it must be gratified by images forming *supplements* rather than *contradictions* of the visible and measurable universe."[6]

For Lovecraft, it is not the sleep of reason that breeds monsters, but reason with its eyes agog. By fusing cutting-edge science with archaic material, Lovecraft creates a twisted materialism in which scientific "progress" returns us to the atavistic abyss, and hard-nosed research revives the factual basis of forgotten and discarded myths. Hence Lovecraft's obsession with archeology; the digs which unearth alien artifacts and bizarrely angled cities are simultaneously historical and imaginal. In the 1930 story "The Whisperer in Darkness," Lovecraft identifies the planet Yuggoth (from which the fungoid Mi-Go launch their clandestine invasions of Earth) with the newly-discovered planet called Pluto. To the 1930 reader – probably the kind of person who

"Azathoth" (Dominique Signoret)

would thrill to popular accounts of C.W. Thompson's discovery of the ninth planet that very year – this factual reference "opens up" Lovecraft's fiction into a real world that is itself opening up to the limitless cosmos.

Lovecraft's most self-conscious, if somewhat strained, fusion of occult folklore and weird science occurs in the 1932 story "The Dreams of the Witch-House." The demonic characters that the folklorist Walter Gilman first glimpses in his nightmares are stock ghoulies: the evil witch crone Keziah Mason, her familiar spirit Brown Jenkin, and a "Black Man" who is perhaps Lovecraft's most unambiguously Satanic figure. These figures eventually invade the real space of Gilman's curiously angled room. But Gilman is also a student of quantum physics, Riemann spaces and non-Euclidian mathematics, and his dreams are almost psychedelic manifestations of his abstract knowledge. Within these "abysses whose material and gravitational properties...he could not even begin to explain," an "indescribably angled" realm of "titan prisms, labyrinths, cube-and-plane clusters and quasi-buildings," Gilman keeps encountering a small polyhedron and a mass of "prolately spheroidal bubbles." By the end of the tale he realizes that these are none other than Keziah and her familiar spirit, classic demonic cliches translated into the most alien dimension of speculative science: hyperspace.

These days, one finds the motif of hyperspace in science fiction, pop cosmology, computer interface design, channelled UFO prophecies, and the postmodern shamanism of today's high-octane psychedelic travellers – all discourses that feed contemporary chaos magic. The term itself was probably coined by the science fiction writer John W. Campbell Jr. in 1931, though its origins as a concept lie in nineteenth-century mathematical explorations of the fourth dimension.

In many ways, however, Lovecraft was the concept's first mythographer. From the perspective of hyperspace, our normal, three-dimensional spaces are exhausted and insufficient constructs. But our incapacity to vividly imagine this new dimension in humanist terms creates a crisis of representation, a crisis that for Lovecraft calls up our most ancient fears of the unknown. "All the objects…were totally beyond description or even comprehension," Lovecraft writes of Gilman's seething nightmare before paradoxically proceeding to describe these horrible objects. In his descriptions, Lovecraft emphasizes the incommensurability of this space through almost non-sensical juxtapositions like "obscene angles" or "wrong" geometry, a rhetorical technique that one Chaos magician calls "Semiotic Angularity."

Lovecraft has a habit of labeling his horrors "indescribable," "nameless, "unseen," "unutterable," "unknown" and "formless." Though superficially weak, this move can also be seen a kind of macabre *via negativa*. Like the apophatic oppositions of negative theologians like Pseudo-Dionysus or St. John of the Cross, Lovecraft marks the limits of language, limits which paradoxically point to the Beyond. For the mystics, this ultimate is the ineffable One, Pseudo-Dionysus' "superluminous gloom" or the *Ain Soph* of the Kabbalists. But there is no unity in Lovecraft's Beyond. It is the omnivorous Outside, the screaming multiplicity of cosmic hyperspace opened up by reason.

For Lovecraft, scientific materialism is the ultimate Faustian bargain, not because it hands us Promethean technology (a man for the eighteenth century, Lovecraft had no interest in gadgetry), but because it leads us beyond the horizon of what our minds can withstand. "The most merciful thing in the world, I think, is the inability of the mind to correlate all its contents," goes the famous opening line of "Call of Cthulhu." By correlating those contents, empiricism opens up "terrifying vistas of reality" – what Lovecraft elsewhere calls "the blind cosmos [that] grinds aimlessly on from nothing to something and from something back to nothing again, neither heeding nor knowing the wishes or existence of the minds that flicker for a second now and then in the darkness".

Lovecraft gave this existentialist dread an imaginative voice, what he called "cosmic alienage". For Fritz Leiber, the "monstrous nuclear chaos" of Azathoth, Lovecraft's supreme entity, symbolizes "the purposeless, mindless, yet all-powerful universe of materialistic belief." But this symbolism isn't the whole story, for, as DMT voyagers know, hyperspace is haunted. The entities that erupt from Lovecraft's inhuman realms seem to suggest that in a blind mechanistic cosmos, the most alien thing is sentience itself. Peering outward through the cracks of domesticated "human" consciousness, a compassionless materialist like Lovecraft could only react with horror, for reason must cower before the most raw and atavistic dream-dragons of the psyche.

Modern humans usually suppress, ignore or constrain these forces lurking in our lizard brain. Mythically, these forces take the form of demons imprisoned under the angelic yokes of altruism, morality, and intellect. Yet if one does not believe in any ultimate universal purpose, then these primal forces are the *most* attuned with the cosmos precisely *because* they are amoral and inhuman. In "The Dunwich Horror", Henry Wheeler overhears a monstrous moan from a diabolical rite and asks "from what unplumbed gulfs of extra-cosmic consciousness or obscure, long-latent heredity, were those half-articular thunder-croakings drawn?" The Outside is within.

Chaos Culture

Lovecraft's fiction expresses a "future primitivism" that finds its most intense esoteric expression in Chaos magic, an eclectic contemporary style of darkside occultism that draws from Thelema, Satanism, Austin Osman Spare, and Eastern metaphysics to construct a thoroughly postmodern magic.

For today's Chaos mages, there is no "tradition". The symbols and myths of countless sects, orders, and faiths, are constructs, useful fictions, "games." That magic works has nothing to do with its truth claims and everything to do with the will and experience of the magician. Recognizing the distinct possibility that we may be adrift in a meaningless mechanical cosmos within which human will and imagination are vaguely comic flukes (the "cosmic indifferentism" Lovecraft himself professed), the mage accepts his groundlessness, embracing the chaotic self-creating void that is himself.

As we find with Lovecraft's fictional cults and grimoires, chaos magicians refuse the hierarchical, symbolic and monotheist biases of traditional esotericism. Like most Chaos magicians, the British occultist Peter Carroll gravitates towards the Black, not because he desires a simple Satanic inversion of Christianity but because he seeks the amoral and shamanic core of magical experience – a core that Lovecraft conjures up with his orgies of drums, guttural chants, and screeching horns. At the same time, Chaos mages like Carroll also plumb the weird science of quantum physics, complexity theory and electronic Prometheanism. Some darkside magicians become consumed by the atavistic forces they unleash or addicted to the dark costume of the Satanic anti-hero. But the most sophisticated adopt a balanced mode of gnostic existentialism that calls all constructs into question while refusing the cold comforts of skeptical reason or suicidal nihilism, a pragmatic and empirical shamanism that resonates as much with Lovecraft's hard-headed materialism as with his horrors.

The first occultist to really engage these notions is Aleister Crowley, who shattered the received vessels of occult tradition while creatively extending the dark dream of magic into the twentieth century. With his outlandish image, trickster texts, and his famous Law of Thelema ("Do what thou wilt shall be the whole of the law"), Crowley called into question the esoteric certainties of "true" revelation and lineage, and was the first magus to give occult antinomionism a decidedly Nietzschean twist.[7]

Unfettered, this occult will to power can easily degenerate into a heartless elitism, and the fascist and racist dimensions of both twentieth-century occultism and Lovecraft himself should not be forgotten. But this self-engendering will is more exuberantly expressed as a will to Art. In many ways, the *fin de siècle* occultism that exploded during Crowley's time was an essentially esthetic esotericism. A good number of the nineteenth-century magicians who inspire us today are the great poets, painters, and writers of Symbolism and decadent Romanticism, many of them dabblers or adepts in Satanism, Rosicrucianism, and hermetic societies. The Hermetic Order of the Golden Dawn was infused with artistic pretensions, and Golden Dawn member and fantasy writer Arthur Machen was one of Lovecraft's strongest influences.

But it was Austin Osman Spare who most decisively dissolved the boundary between artistic and magical life. Though working independently of the Surrealists, Spare also based his art on the dark and autonomous eruptions of "subconscious" material, though in a more overtly theurgic context.[8] Today's Chaos magicians are heavily influenced by Spare, and their Lovecraftian rites express this simultaneously creative and nihilistic dissolution. And as postmodern spawn of role-playing games, computers, and pop culture, they celebrate the fact that Lovecraft's secrets are scraped from the barrel of pulp fiction.

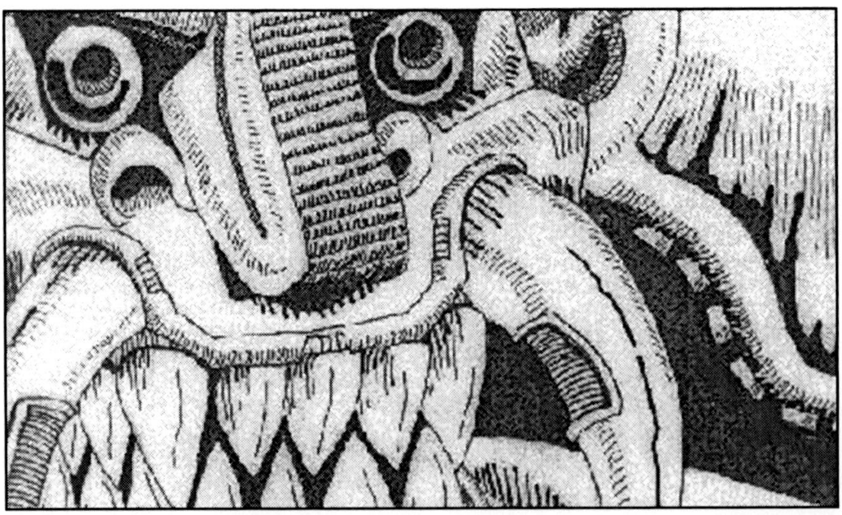

Proof in the Pudding

In a message cross-posted to the Internet newsgroups alt.necromicon [*sic*] and alt.satanism, Parker Ryan listed a wide variety of magical techniques described by Lovecraft, including entheogens, glossalalia, and shamanic drumming. Insisting that his post was "not a satirical article," Ryan then described specific Lovecraftian rites he had developed, including this "Rite of Cthulhu":

> A) Chanting. The use of the "Cthulhu chant" to create a concentrative or meditative state of consciousness that forms the basis of much later magickal work.
>
> B) Dream work. Specific techniques of controlled dreaming that are used to establish contact with Cthulhu.
>
> C) Abandonment. Specific techniques to free oneself from culturally conditioned reality tunnels.

Ryan goes on to say that he's experimented with most of his rites "with fairly good success."

In coming to terms with the "real magic" embedded in Lovecraft, one quickly encounters a fundamental irony: the cold skepticism of Lovecraft himself. In his letters, Lovecraft poked fun at his own tales, claiming he wrote them for cash and playfully naming his monsters after his friends. While such attitudes in no way diminish the imaginative power of Lovecraft's tales – which, as always, lie outside the control and intention of their author – they do pose a problem for the working occultist seeking to establish Lovecraft's magical authority.

The most obvious, and least interesting, answer is to find authentic magic in Lovecraft's biography. Lovecraft's father was a traveling salesman who died in a madhouse when Lovecraft was eight, and vague rumors that he was an initiate in some Masonic order or other were exploited in the *Necronomicon* cobbled together by George Hay, Colin Wilson, and Robert Turner. Others have tried to track Lovecraft's occult know-how, especially his familiarity with Aleister Crowley and the Golden Dawn. In an Internet document relating the history of the "real" *Necronomicon*, Colin Low argues that Crowley befriended Sonia Greene in New York a few years before the woman married Lovecraft. As proof of Crowley's indirect influence on Lovecraft, Low sites this intriguing passage from "The Call of Cthulhu":

> That cult would never die until the stars came right again and the secret priests would take Cthulhu from His tomb to revive His subjects and resume His rule of earth. The time would be easy to know, for then mankind would have become as the Great Old Ones; free and wild, and beyond good and evil, with laws and morals thrown aside and all men shouting and killing and revelling in joy. Then the liberated Old Ones would teach them new ways to shout and kill and revel and enjoy themselves, and all earth would flame with a holocaust of ecstasy and freedom.

Low claims this passage is a mangled reflection of Crowley's teachings on the new Aeon and the *The Book of the Law*. In an article in *Societé*, Robert North also states that Lovecraft referred to "A.C." in a letter, and that Crowley was mentioned in Leonard Cline's *The Dark Chamber*, a novel Lovecraft discussed in his *Supernatural Horror in Literature.*

But so what? Lovecraft was a fanatical and imaginative reader, and many such folks are drawn to the semiotic exotica of esoteric lore regardless of any beliefs in or experiences of the paranormal. From *The Case of Charles Dexter Ward* and elsewhere, it's clear that Lovecraft knew the basic outlines of the occult. But these influences pale next to *Vathek*, Poe, or Lord Dunsany.

Desperate to assimilate Lovecraft into a "tradition", some occultists enter into dubious explanations of mystical influence by disincarnate beings. North gives this Invisible College idea a shamanic twist, asserting that prehistoric Atlantean tribes who survived the flood exercised telepathic influence on people like John Dee, Blavatsky, and Lovecraft. But none of these Lovecraft hierophants can match the delirious splendor of Kenneth Grant. In *The Magical Revival,* Grant points out more curious similarities between Lovecraft and Crowley: both refer to "Great Old Ones" and "Cold Wastes" (of Kadath and Hadith, respectively); the entity "Yog-Sothoth" rhymes with "Set-Thoth," and *Al Azif: The Book of the Arab* resembles Crowley's *Al vel Legis: The Book of the Law.* In *Nightside of Eden*, Grant maps Lovecraft's pantheon onto a darkside Tree of Life, comparing the mangled "iridescent globes" that occasionally pop up in Lovecraft's tales with the shattered *sefirot* known as the Qlipoth. Grant concludes that Lovecraft had "direct and conscious experience of the inner planes,"[9] the same zones Crowley prowled, and that Lovecraft "disguised" his occult experiences as fiction.

Like many latter-day Lovecraftians, Grant commits the error of literalizing a purposefully nebulous myth. A subtler and more satisfying version of this argument is the notion that Lovecraft had direct *unconscious* experiences of the inner planes, experiences which his quotidian mind rejected but which found their way into his writings nonetheless. For Lovecraft was blessed with a vivid and nightmarish dream life, and drew the substance of a number of his tales from beyond the wall of sleep.

In this sense Lovecraft's magickal authority is nothing more or less than the authority of dream. But what kind of dream tales are these? A Freudian could have a field day with Lovecraft's fecund, squishy sea monsters, and a Jungian analyst might recognize the liniments of the proverbial shadow. But Lovecraft's shadow is so inky it swallows the standard archetypes of the collective unconscious like a black hole. If we see the archetypal world not as a static storehouse of timeless godforms but as a constantly mutating carnival of figures, then the seething extraterrestrial monsters that Lovecraft glimpsed in the chaos of hyperspace are not so much archaic figures of heredity than the avatars of a new psychological and mythic aeon. At the very least, it would seem that things are getting mighty out of hand beyond the magic circle of the ordered daylight mind.

In an intriguing Internet document devoted to the *Necronomicon*, Tyagi Nagasiva places Lovecraft's potent dreamtales within the *terma* tradition found in the Nyingma branch of Tibetan Buddhism.[10] *Termas* were "pre-mature" writings hidden by Buddhist sages for centuries until the time was ripe, at which point religious visionaries would divine their physical hiding places through omens or dreams. But some *termas* were revealed entirely in dreams, often couched in otherworldly Dakini scripts. An old Indian revisionary tactic (the second-century Nagarjuna was said to have discovered his Mahayana masterpieces in the serpent realm of the *nagas*), the *terma* game resolves the religious problem of how to alter a tradition without disrupting traditional authority. The famous *Tibetan Book of the Dead* is a *terma*, and so, perhaps, is the *Necronomicon*.

Of course, for Chaos magicians, reality can coherently present itself through any number of self-sustaining but mutually contradictory symbolic paradigms (or "reality tunnels," in Robert Anton Wilson's memorable phrase). Nothing is true and everything is permitted. By emphasizing the self-fulfilling nature of all reality claims, this postmodern perspective creatively erodes the distinction between legitimate esoteric transmission and total fiction.

This bias toward the experimental is found in Anton LaVey's *Satanic Rituals*, which includes the first overtly Lovecraftian rituals to see print. In presenting "*Die Elektrischen Vorspiele*" (which LaVey based on a Lovecraftian tale by Frank Belknap Long), the "Ceremony of the Angles," and "The Call to Cthulhu" (the latter two penned by Michael Aquino), LaVey does claim that Lovecraft "clearly...had been influenced by very real sources."[11]

But in holding that Satanic magic allows you to "objectively enter into a subjective state," LaVey more emphatically emphasizes the ritual power of fantasy – a radical subjectivity which explains his irreverence towards occult source material, whether Lovecraft or Masonry. In naming his Order of the Trapezoid after the "Shining Trapezohedron" found in Lovecraft's "The Haunter of the Dark" – a black, oddly-angled extraterrestrial crystal used to communicate with the Old Ones – LaVey emphasized that fictions can channel magical forces regardless of their historical authenticity.

In his two rituals, Michael Aquino expresses the subjective power of "meaningless" language by creating a "Yuggothic" tongue similar to that heard in Lovecraft's "The Dunwich Horror" and "The Whisperer in the Dark." Such guttural utterances help to shut down the rational mind (try chanting "P'garn'h v'glyzz" for a couple of hours), a notion elaborated by Kenneth Grant in his notion of the Cult of Barbarous Names. After leaving the Church of Satan to form the more serious Temple of Set in 1975, Aquino eventually reformed the Order of the Trapezoid into the practical magic wing of the Setian philosophy. For Stephen R. Flowers, once the Grand Master of the order, the substance of Lovecraftian magic is precisely an overwhelming subjectivity that flies in the face of objective law. "The Old Ones are the objective manifestations…of the subjective universe which is what is trying to 'break through' the merely rational mind-set of modern humanity."[12] For Flowers, such invocations are ultimately apocalyptic, hastening a transition into a chaotic aeon in which the Old Ones reveal themselves as future reflections of the Black Magician ("There are no more Nightmares for us," he wrote me).

This desire to rebel against the tyranny of reason and its ordered objective universe is one of the underlying goals of Chaos magic. Many would applaud the sentiment expressed by Albert Wilmarth in Lovecraft's "The Whisperer in Darkness":

> To shake off the maddening and wearying limitations of time and space and natural law – to be linked with the vast *outside* – to come close to the nighted and abysmal secrets of the infinite and ultimate – surely such a thing was worth the risk of one's life, soul, and sanity![13]

In his electronically circulated text "Kathulu Majik: Luvkrafting the Roles of Modern Uccultizm," Tyagi Nagasiva ('Haramullah') writes that most Western magic is ossified and dualistic, heavily weighted towards the forces of order, hierarchy, moralizing, and structured language. "Without the destabilizing force of Kaos, we would stagnate intellectually, psychologically and otherwise…Kathulu provides a necessary instability to combat the stolid and fixed methods of the structured 'Ordurs'…One may become balanced through exposure to Kathulu" (Tyagi's "mis-spellings" show the influence of Genesis P. Orridge's Temple of Psychick Youth). 'Haramullah' criticizes black magicians who simply reverse "Ordur" with "Kaos," rather than bringing this underlying polarity into balance (a dualistic error he also finds in Lovecraft). Showing strong Taoist and Buddhist influences, Haramullah calls instead for a "Midul Path" that magically navigates between structure and disintegration, will and void. "The idea that one may progress linearly along the MP [Midul Path] is mistaken. One becomes, one does not progress. One attunes, one does not forge. One allows, one does not make."

In the *Cincinatti Journal of Ceremonial Magic,* the anonymous author of "Return of the Elder Gods" presents an evolutionary reason for Mythos magic. The author accepts the scenario of an approaching world crisis brought on by the invasion of the Elder Gods, Qlipothic transdimensional entities who ruled protohumanity until they were banished by "the agent of the Intelligence," a Promethean figure who set humanity on its current course of evolution. We remain connected to these Elder Gods through the "Forgotten Ones," the atavistic forces of hunger, sex and violence that linger in the subterranean levels of our being. Only by magically "reabsorbing" the Forgotten Ones and using the subsequent energy to bootstrap higher consciousness can we keep the portal sealed against the return of the Elder Gods. Though Lovecraft's name is never mentioned in the article, he is ever present, a skeptical materialist dreaming the dragons awake.

Writing the Dream...

Within the Mythos tales, one finds two dimensions – the normal human world and the infested Outside – and it's the ontological tension between them that powers Lovecraft's magick realism. Though Cthulhu and friends have material aspects, their reality is most horrible for what it says about the way the universe *is*. As the Lovecraft scholar S.T. Joshi notes, Lovecraft's narrators frequently go mad "not through any physical violence at the hands of supernatural entities but through the mere *realization* of the the existence of such a race of gods and beings." Faced with "realms whose mere existence stuns the brain," they experience severe cognitive dissonance – precisely the sorts of disorienting rupture sought by Chaos magicians.[14]

The role-playing game *Call of Cthulhu* wonderfully expresses the violence of this Lovecraftian paradigm shift. In adventure games like *Dungeons & Dragons*, one of your character's most significant measures is its hit points – a number which determines the amount of physical punishment your character can take before it gets injured or dies. *Call of Cthulhu* replaces this physical characteristic with the psychic category of Sanity. Face-to-face encounters with Yog-Sothoth or the insects from Shaggai knock points off your Sanity, but so does your discovery of more information about the Mythos – the more you find out from books or starcharts, the more likely you are to wind up in the Arkham Asylum. Magic also comes with an ironic price, one that Lovecraftian magicians might well pay heed to. If you use any of the binding spells from *De Vermis Mysteriis* or the

"Shub-Niggurath" (Dominique Signoret)

Pnakotic Manuscripts, you necessarily learn more about the Mythos and thereby lose more sanity.[15]

Lovecraft's scholarly heroes also investigate the Mythos as much through reading and thinking as through movements through physical space, and this psychological exploration draws the mind of the reader directly into the loop. Usually, readers suspect the dark truth of the Mythos while the narrator still clings to a quotidian attitude – a technique that subtly forces the reader to identify with the Outside rather than with the conventional worldview of the protagonist. Magically, the blindness of Lovecraft's heroes corresponds to a crucial element of occult theory developed by Austin Osman Spare: that magic occurs over and against the conscious mind, that ordinary thinking must be silenced, distracted, or thoroughly deranged for the chthonic will to express itself.[16]

In order to invade our plane, Lovecraft's entities need a portal, an interface between the worlds, and Lovecraft emphasizes two: books and dreams. In "Dreams of the Witch-House," "The Shadow out of Time" and "The Shadow over Innsmouth," dreams infect their hosts with a virulence that resembles the more overt psychic possessions that occur in "The Haunter in the Dark" and *The Case of Charles Dexter Ward*. Like the monsters themselves, Lovecraft's dreams are autonomous forces breaking through from Outside and engendering their own reality.

But these dreams also conjure up a more literal "outside": the strange dream life of Lovecraft himself, a life that (as the informed fan knows) directly inspired some of the tales.[17] By seeding his texts with his own nightmares, Lovecraft creates a autobiographical homology between himself and his protagonists. The stories themselves start to dream, which means that the reader too lies right in the path of the infection.

Lovecraft reproduces himself in his tales in a number of ways – the first-person protagonists reflect aspects of his own reclusive and bookish lifestyle; the epistolary form of the "The Whisperer in Darkness" echoes his own commitment to regular correspondence;

character names are lifted from friends; and the New England landscape is his own. This psychic self-reflection partially explains why Lovecraft fans usually become fascinated with the man himself, a gaunt and solitary recluse who socialized through the mail, yearned for the eighteenth century, and adopted the crabby outlook and mannerisms of an old man. Lovecraft's life, and certainly his voluminous personal correspondence, form part of his myth.

Lovecraft thus solidifies his virtual reality by adding autobiographical elements to his shared world of creatures, books and maps. He also constructs a documentary texture by thickening his tales with manuscripts, newspaper clippings, scholarly citations, diary entries, letters, and bibliographies that list fake books alongside real classics. All this produces the sense that "outside" each individual tale lies a meta-fictional world that hovers on the edge of our own, a world that, like the monsters themselves, is constantly trying to break through and actualize itself. And thanks to Mythos storytellers, role-playing games, and dark-side magicians, it has.

...And Dreaming the Book

In "The Shadow out of Time," Lovecraft makes explicit one of the fantastic equations that drives his Magick Realism: the equivalence of dreams and books. For five years, the narrator, an economics professor named Nathaniel Wingate Peaslee, is taken over by a mysterious "secondary personality." After recovering his original identity, Peaslee is beset by powerful dreams in which he finds himself in a strange city, inhabiting a huge tentacle-sprouting conical body, writing down the history of the modern Western world in a book. In the climax of the tale, Peaslee journeys to the Australian desert to explore ancient ruins buried beneath the sands. There he discovers a book written in English, in his own handwriting: the very same volume he had produced inside his monstrous dream body.

Though we learn very little of their contents, Lovecraft's diabolical grimoires are so infectious that even glancing at their ominous sigils proves dangerous. As with their dreams, these texts obsess Lovecraft's bookish protagonists to the point that the volumes, in Christopher Frayling's phrase, "vampirize the reader." Their titles alone are magic spells, the hallucinatory incantations of an eccentric antiquarian: the *Pnakotic Manuscripts*, the *Ilarnet Papyri*, the *R'lyeh Text*, the *Seven Cryptical Books of Hsan*. Lovecraft's friends contributed *De Vermis Mysteriis* and von Junzt's *Unaussprechlichen Kulten*, and Lovecraft named the author of his *Cultes Des Goules*, the Comte d'Erlette, after his young fan August Derleth. Hovering over all these grim tomes is the "dreaded" and "forbidden" *Necronomicon*, a book of blasphemous invocations to speed the return of the Old Ones. Lovecraft's supreme intertextual fetish, the *Necronomicon* stands as one of the few mythical books in literature that have absorbed so much imaginative attention that they've entered published reality.

If books owe their life not to their individual contents but to the larger intertextual webwork of reference and citation within which they are woven, then the dread *Necronomicon* clearly has a life of its own. Besides literary studies, the *Necronomicon* has generated numerous pseudo-scholarly analyses, including significant appendixes in the *Encyclopedia Cthulhiana* and Lovecraft's own "History of the *Necronomicon*." A number of FAQs can be found on the Internet, where a mild flame war periodically erupts between magicians, horror fans, and mythology experts over the reality of the book. The undead entity referred to in the *Necronomicon's* famous couplet – "That is not dead which can eternal lie / And with strange eons even death may die" – may be nothing more or less than the the text itself, always lurking in the margins as we read the real.

Lovecraft's brief "History" was apparently inspired by the first *Necronomicon* hoax: a review of an edition of the dreaded tome submitted to Massachusetts' *Branford Review* in 1934.[18] Decades later, index cards for the book started popping up in university library catalogs.

It is perhaps the principle expression of Lovecraft's Magick Realism that all these ghostly references would finally manifest the book itself. In 1973, a small-press edition of *Al Azif* (the *Necronomicon's* Arabic name) appeared, consisting of eight pages of simulated Syrian script repeated 24 times. Four years later, the Satanists at New York's Magickal Childe published a *Necronomicon* by Simon, a grab bag that contains far more Sumerian myth than Lovecraft (though portions were "purposely left out" for the "safety of the reader"). George Hay's *Necronomicon: The Book of Dead Names,* also a child of the '70s, is the most complex, intriguing, and Lovecraftian of the lot. In the spirit of the master's pseudoscholarship, Hay nests the fabulated invocations of Yog-Sothoth and Cthulhu amongst a set of analytic, literary and historical essays.

Though magicians with strong imaginations have claimed that even the Simon book works wonders, the pseudohistories of the various *Necronomicons* are far more compelling than the texts themselves. Lovecraft himself provided the bare bones: the text was penned in 730 A.D by a poet, the Mad Arab Abdul Alhazred, and named after the nocturnal sounds of insects. It was subsequently translated by Theodorus Philetas into Greek, by Olaus Wormius into Latin, and by John Dee into English. Lovecraft lists various libraries and private collections where fragments of the volume reside, and gives us a knowing wink by noting that the fantasy writer R.W. Chambers is said to have derived the monstrous and suppressed book found in his novel *The King in Yellow* from rumors of the *Necronomicon* (Lovecraft himself claimed to have gotten his inspiration from Chambers).

All of the *Necronomicon's* subsequent pseudohistories weave the book in and out of actual occult history, with John Dee playing a particularly conspicuous role. According to Colin Wilson, the version of the text published in the Hay *Necronomicon* was encrypted in Dee's

Enochian cipher-text *Liber Logoaeth*. Colin Low's *Necronomicon* FAQ claims that Dee discovered the book at the court of King Rudolph II's court in Prague, and that is was under its influence that Dee and his scryer Edward Kelly achieved their most powerful astral encounters. Never published, Dee's translation became part of celebrated collection of Elias Ashmole housed at the British Library. Here Crowley read it, freely cobbling passages for *The Book of the Law*, and ultimately passing on some of its contents indirectly to Lovecraft through Sophia Greene. Crowley's role in Low's tale is appropriate, for Crowley certainly knew the magical power of hoax and history.

For the history of the occult *is* a confabulation, its lies wedded to its genealogies, its "timeless" truths fabricated by revisionists, madmen, and geniuses, its esoteric traditions a constantly shifting conspiracy of influences. The *Necronomicon* is not the first fiction to generate real magical activity within this potent twilight zone between philology and fantasy.

To take an example from an earlier era, the anonymous Rosicrucian manifestos that first appeared in the early 1600s claimed to issue from a secret brotherhood of Christian Hermeticists who finally deemed it time to come above ground. Many readers immediately wanted to join up, though it is unlikely that such a group existed at the time. But this hoax focused esoteric desire and inspired an explosion of "real" Rosicrucian groups. Though one of the two suspected authors of the manifestos, Johann Valentin Andreae, never came clean, he made veiled references to Rosicrucianism as an "ingenius game which a masked person might like to play upon the literary scene, especially in an age infatuated with everything unusual."[19] Like the Rosicrucian manifestos or Blavatsky's *Book of Dzyan*, Lovecraft's *Necronomicon* is the occult equivalent of Orson Welles' radio broadcast of the "War of the Worlds." As Lovecraft himself wrote, "No weird story can truly produce terror unless it is devised with all the care and verisimilitude of an actual hoax."[20]

In *Foucault's Pendulum*, Umberto Eco suggests that esoteric truth is perhaps nothing more than a semiotic conspiracy theory born of an endlessly rehashed and self-referential literature – the intertextual fabric Lovecraft understood so well. For those who need to ground their profound states of consciousness in objective correlatives, this is a damning indictment of "tradition." But as Chaos magicians remind us, magic is nothing more than subjective experience interacting with an internally consistent matrix of signs and affects. In the absence of orthodoxy, all we have is the dynamic tantra of text and perception, of reading and dream. These days the Great Work may be nothing more or less than this "ingenius game," fabricating itself without closure or rest, weaving itself out of the resplendent void where Azathoth writhes on his Mandelbrot throne.

Erik Davis is the author of the forthcoming collection *Nomad Codes: Adventures in Modern Esoterica*; he also penned *The Visionary State: A Journey through California's Spiritual Landscape*, the cult classic *Techgnosis: Myth, Magic, and Mysticism in the Information Age*, and a critical volume about magick and Led Zeppelin. A frequent speaker and teacher at universities and festivals alike, Davis has contributed to dozens of books and journals. He also hosts the weekly net radio show Expanding Mind on the Progressive Radio Network, and posts at www.techgnosis.com. He is currently pursuing a PhD at Rice University.

How Martin Gardner Bamboozled the Skeptics

A lesson in trusting a magician

by *Greg Taylor*

The modern skeptical 'movement' has grown and thrived in recent years to the point where the public generally views self-appointed 'skeptics' as arbiters of the truth and defenders of rational thought. But how much of what they say can we really trust as being objective truth? Are self-described skeptics championing critical thinking, or are they simply defending one particular worldview? The late Marcello Truzzi came to think so: despite being the co-founding chairman of the influential skeptical group CSICOP (the Committee for the Scientific Investigation of Claims of the Paranormal), Truzzi soon became disillusioned with the organization, saying they "tend to block honest inquiry, in my opinion… Most of them are not agnostic toward claims of the paranormal; they are out to knock them." Truzzi claimed that by using the title of 'skeptic', biased debunkers had claimed an authority that they were not entitled to, opining that "critics who take the negative rather than an agnostic position but still call themselves 'skeptics' are actually pseudo-skeptics and have, I believed, gained a false advantage by usurping that label."[1] Should we be more skeptical of the skeptics?

If there is one skeptic who stands above all others in terms of being regarded as an authoritative voice, it must surely be Martin Gardner. Through the course of his life, Gardner – who passed away aged 95 in May 2010 – published more than seventy books on such diverse topics as mathematics, science, philosophy, literature and skepticism. For a quarter of a century he was also the writer of the 'Mathematical Games' column in *Scientific American*, and as a consequence he has influenced many of the modern day's top academics in the hard sciences. His opinion therefore commands much respect from intellectuals. Every two years a 'Gathering for Gardner' is held to celebrate his lifelong contributions (to maths in particular), and has been attended by the likes of Stephen Wolfram and John Conway. Douglas Hofstadter described Gardner as "one of the great intellects produced in this country in this century," and Arthur C. Clarke once labeled him a "national treasure."

Gardner has also long been one of the major voices in the skeptical movement; George Hansen describes him as "the single most powerful critic of the paranormal in the second half of the 20th century". Gardner was writing 'skeptical' books long before the modern movement 'began' in earnest with the inception of CSICOP (now known as CSI) in the 1970s – his seminal deconstruction of pseudoscience, *In the Name of Science* (later renamed *Fads and Fallacies in the Name of Science*), had been published two decades previous in 1952. Gardner's antipathy towards the supernatural was perhaps a natural outgrowth of his academic interests, skepticism, proficiency and understanding of conjuring techniques, and – improbably – his religious beliefs. Gardner was not a Christian, but he did believe in

God; a primary motivation for his criticisms of parapsychology might be found in an essay he wrote titled "Prayer: Why I Do Not Think It Foolish", in which he says "I am among those theists who, in the spirit of Jesus' remark that only the faithless look for signs, consider such tests both futile and blasphemous... Let us not tempt God".[2] Nevertheless, Gardner's esteemed standing amongst academics has allowed his skeptical writings to be widely accepted as the final word on controversial topics. In the words of Stephen Jay Gould: "For more than half a century, Martin Gardner has been the single brightest beacon defending rationality and good science against the mysticism and anti-intellectualism that surrounds us."

As an example of Gardner's influence on discourse about paranormal topics, consider the reference to one of his essays in the *New York Times* review of Deborah Blum's book *Ghost Hunters: William James and the Search for Scientific Proof of Life After Death.* Blum's book tells the (partial) story of the Society for Psychical Research (SPR) – a group of academics, including the likes of William James, Sir William Crookes and Sir Oliver Lodge, who came together in the

Martin Gardner
(Konrad Jacobs, Erlangen. Creative Commons Share-Alike Attribution Licence)

late 19th century to try to solve the question of life after death – and its investigation of Mrs. Leonora Piper, one of the most successful mediums of modern times. Anthony Gottlieb's *New York Times* review suggests that Blum might have taken a different view of Mrs. Piper's mediumship if she had read Martin Gardner's critique of the SPR's research prior to writing *Ghost Hunters*:

> The book is peppered with narratives reporting ghost stories and seances. Blum writes that these are "derived from" documents, often from the society's archives, although the telling of them is her own. But these narratives obscure the methods that mediums like Mrs. Piper used – methods that have been explained by debunkers like Martin Gardner, who in 1992 published a long exposé called "How Mrs. Piper Bamboozled William James." For example, Blum's ghost narratives do not show, as Gardner did, how Mrs. Piper fished for information by gauging her sitters' responses to all her wrong answers, or mined the information available from earlier sittings, from sittings with others and from things said while her sitters believed she was unconscious in a trance.[3]

In this particular instance all of Blum's narratives, *sourced from documents written by the original researchers*, are painted over with one broad brush-stroke by invoking Gardner's name. His essay, it seems, lays bare the techniques of deception used by Mrs. Piper – a mix of fishing (also known as cold reading) and devious information mining – which have over the course of a century fooled researchers and writers lacking the critical thinking skills and magical insights of a Martin Gardner. This is a bold claim – the original SPR reports on the Piper case had contributions from some of the finest minds of the time, and run to thousands of pages in total length. In short, the investigation of the mediumship of Mrs. Piper is one of the most comprehensive and well-documented in the history of psychical research. Is Gardner's authority in this case being overstated?

Sleight of Pen

Leonora Piper is considered by many to be the foremost 'trance medium' in the history of psychical research. She was investigated by the SPR over a period of more than 40 years, beginning in the mid-1880s, and convinced nearly all researchers that she had access to information that was not available by 'normal' means. During her séances Mrs. Piper would fall into a trance, and her body would be 'taken over' by another consciousness known as a control (who often claimed to be the spirit of a deceased individual). This control would sometimes communicate directly with the sitter, and at other times would mediate communication from other spirits wishing to talk to the sitter. For the first 8 years of the SPR's investigation her main control was 'Phinuit', a deceased French physician (though the SPR never found any historical trace of his identity), who spoke to sitters using Mrs. Piper's voice. After Phinuit came 'G.P', a writer named George Pellew who was known to the SPR whilst alive, who died in an accident in 1892. During his 'tenure' communication continued via Mrs. Piper's voice, but also with the addition of writing as well. In later years, a number of 'advanced spirits' known as the Imperator Group took over control of 'the machine' (their name for Mrs. Piper), with nearly all communication being via writing. The long period of the Piper investigation even resulted in a number of the original researchers 'returning' as spirits communicating from 'the other side'.

Martin Gardner's article on Mrs. Piper first appeared in the 1996 book *The Encyclopedia of the Paranormal*, and in 2003 was included in an anthology of his writings titled *Are Universes Thicker Than Blackberries?* However, it is based on a longer essay published in *Free Inquiry* in 1992 (more simply titled "Communicating with the dead: William James and Mrs. Piper"), which was later included in another anthology of his writings, *The Night is Large*. Gardner prefaces this earlier essay with a statement of his "unbounded" admiration for William James, followed by what he thought was the reason for James's "gullibility with respect to psychic phenomena":

> Had James been better informed about techniques of deception, practiced by magicians and mediums, he would not have been so impressed by Mrs. Piper's carefully contrived persona. Moreover, James had only a weak comprehension of how to conduct controlled tests of mediums.[4]

This criticism – that Mrs. Piper's positive results would have been less impressive with some knowledge of "techniques of deception" – is the first to set off alarm bells that Gardner may be dealing cards from the bottom of the deck when it comes to presenting the facts of this particular case. Although it glosses over the fact that William James had enough wits about him to debunk other fraudulent mediums before Piper, there is a far more important omission. Martin Gardner focuses almost exclusively on William James in his essay on the mediumship of Mrs. Piper, despite the fact that the primary researcher in this case was not James, but Richard Hodgson (James 'discovered' Piper in 1885 and spent some time investigating her on behalf of the SPR; Hodgson took over the investigation in 1887, and continued as the lead researcher in the United States until his death in 1905). Gardner's criticism seems a whole lot less impressive once one knows that Richard Hodgson began his investigation highly skeptical of paranormal claims (though equally open to genuine evidence); he once warned others interested in Spiritualism that "nearly all the professional mediums are a gang of vulgar tricksters who are more or less in league with one another." Hodgson was also extremely well-versed in the trickery employed by fraudulent mediums – he is famously known for a brutal exposé of the chicanery of the founder of Theosophy, Madame Blavatsky, and with S.J. Davy constructed a series of James Randi-like fake séances to test the observational

Richard Hodgson

abilities of sitters and researchers, the results of which were reported in the *Proceedings of the SPR* under the title "The Possibilities of Malobservation and Lapse of Memory from a Practical Point of View".[5] Richard Hodgson *was* well-informed about techniques of deception, and he investigated Piper for 18 years on behalf of the SPR. And yet in Gardner's essay he is mentioned only in passing, and then only as "a British psi researcher" (though Hodgson was actually an Australian) "who came to Boston in 1887 to serve as secretary of the American Society for Psychical Research (ASPR), and to edit its journal."

Did Gardner deliberately omit this important point? Or was he so lacking in knowledge of the source material that he didn't even know who the primary investigator in the Piper case was? The latter conclusion may be supported by another odd aspect of Gardner's essay: he gets dates wrong. He says that Richard Hodgson died in 1909, when he passed away in 1905. Similarly, Gardner claims that the writer George Pellew died in 1881, when he died in 1892. It could certainly be the case that these mistakes were not the author's failing…many an editor or typesetter has committed errors as a manuscript passed through their hands on the way to publication.[6] However, it is less easy to forgive other dating errors which can only be attributed to the author. For instance, Gardner asserts that "when [the alleged spirit of] Pellew began coming through Mrs. Piper, Hodgson was so skeptical that he hired detectives to shadow Mrs. Piper and her husband for several weeks to make sure they were not researching information about his friend." In fact, Hodgson hired detectives almost immediately after he took on the Piper case, around 1887-88, whereas Pellew didn't die and start 'coming through' until 1892. Anybody who has read the original reports of the Piper case would know that Gardner is wrong in this instance – Fred Myers mentions Hodgson's use of detectives in his 1890 introduction to 'the British report' on Mrs. Piper's mediumship in the *Proceedings of the SPR*.[7]

Furthermore Gardner states ‘facts’ which are not just debatable, but completely the opposite to what can be found in the primary sources. He says of Mrs. Piper’s trances that they “never occurred spontaneously”, despite Fred Myers stating in his report that “[T]he trance has occasionally appeared when it was not desired.”[8] Also, says Gardner, “they never began when she was alone or asleep”. Again, from the same page of the same primary source: “the access has several times come upon her during sleep.”[9] Further, according to Gardner, “whenever a sitter paid for a séance, she had no difficulty going into a trance.” At this point we might assume that Gardner wasn’t too familiar with Fred Myers’ report on Mrs. Piper, as it continues (yet again on the same page): “These trances cannot always be induced at pleasure. A state of quiet expectancy or ‘self-suggestion’ will usually bring one on; but sometimes the attempt altogether fails.”[10] And in case Gardner missed Myers’ statements, a rebuttal can be also be found on page 1(!) of probably the most important primary source concerning Mrs. Piper’s mediumship, Richard Hodgson’s report in Volume VIII of the *Proceedings of the SPR*: “Several times Mrs. Piper was unable to go into trance at all.”[11]

There are additional reasons for thinking that Gardner may not have consulted the original research. For instance, he asserts that records of Mrs. Piper’s séances “show plainly that her controls did an enormous amount of what was called ‘fishing’, and today is called ‘cold reading’ [a technique where the ‘medium’ starts with educated guesses and then narrows in on only the positive responses from the sitter]”. Later he boldly says “cunning cold reading may account for most of Mrs. Piper’s hits.” It is important to note that Gardner is not offering a new insight and explanation for Mrs. Piper’s success here, despite what reviewers for the *New York Times* believe – he is in fact directly disagreeing with the considered opinion of the original investigators, who referenced the possibility of fishing/cold reading numerous times and dismissed it as an explanation. For example, James Hyslop said he “applied fishing, guessing, shrewd inference, and suggestion and found them wanting.”[12] Professor William Newbold noted that both he and Richard Hodgson "have seen much of professional mediums, and are thoroughly familiar with the methods of 'fishing' upon which they generally rely. Hence we always had such possibilities in mind, and it would have been impossible for any large amount of detailed information to have been extracted from us in this way without our knowledge."[13] Contra the claims of Martin Gardner, Frank Podmore – perhaps the most cynical of all the SPR investigators – noted the hypothesis of cold reading was "clearly inadequate to explain even a small fraction of the later records".[14] Podmore noted that initially the investigators were wary that Mrs. Piper was using cold reading because...

> ...both Christian and surnames emerge frequently piecemeal, and obviously with considerable effort. This tentative exhibition of important information was naturally regarded at the earlier sittings as a suspicious circumstance, pointing to "fishing"... But in so far as it is founded on this circumstance, that reputation is apparently undeserved. For this same tentative and piecemeal emergence of proper names appears in the most fully reported sittings, where it is tolerably certain that no hints were given, and even in those sittings where the communication is made entirely by writing.[15]

Gardner also fails to note that a number of successful 'hits' by Mrs. Piper were on details which were not known to the sitters – that is, where the person that Mrs. Piper was supposedly 'cold reading' did not know anything of the details being 'fished'.

These points are of crucial importance in judging Gardner's essay, because – as the title notes – Gardner is supposedly showing "how Mrs. Piper bamboozled William James." He writes that "James was aware of how Mrs. Piper's controls shamelessly fished for data, yet he could not avoid thinking her messages were 'accreted round some originally genuine nucleus'." There is actually a very good reason why James looked elsewhere than the 'fishing' hypothesis for an explanation – the original investigators considered it, and dismissed it. The uninformed reader would assume that Gardner has seen something that a naïve William James did not, when the reality of the situation is quite the opposite.

Being a knowledgeable magician, Gardner knows that there are other possible techniques for achieving 'psychic abilities', and so he doesn't just rest with the fishing claim (even if he ascribes "most of Mrs. Piper's hits" to it). For instance, he also points out that "Mrs. Piper usually held a client's hand throughout a sitting", which could be suggestive of 'muscle reading'. This is certainly a point worth bringing up when investigating claims of mediumship, as it is another magician's trick for 'reading the mind' and first came to prominence during the 1870s,[16] little more than a decade before the testing of Mrs. Piper began. Muscle reading takes advantage of the ideomotor effect, where very slight involuntary reactions to questions can be ascertained through physical contact, often by holding a person's hand. Ironically, William James was one of the first to bring the ideomotor effect to wide attention, in his foundational textbook *Principles of Psychology* – a point Martin Gardner does not mention in his essay. More concerning though is that once again Gardner seems to be disregarding the source material. For instance, in the British report on Mrs. Piper, Sir Oliver Lodge wrote:

> I am familiar with muscle-reading and other simulated 'thought-transference' methods, and prefer to avoid contact whenever it is possible to get rid of it without too much fuss. Although Mrs. Piper always held somebody's hand while preparing to go into the trance, she did not always continue to hold it when speaking as Phinuit.[17]

It is also worth noting that for the bulk of her career, Mrs. Piper's 'communicators' used her hand(s) to write, rather than speaking 'through' her voice, severely limiting any chance of contact mind reading. It is difficult to see how Gardner translates Lodge's notes on avoiding contact, and the reported circumstances of Mrs. Piper's trance, into her "*usually* holding a client's hand *throughout* a sitting" [my emphasis]. Similarly, elsewhere in his essay he claims that Mrs.

Piper's eyes "were often only half closed, allowing her to observe reactions". Gardner seems to ignore the fact that for the majority of her sittings, Piper's trance conditions had her head buried in a pillow, with the eyes "not only closed but turned from the sitter."[18]

Not content with fishing, muscle reading, and visual cues (though all dismissed by the original researchers), Gardner believes…

> …she had other tricks up her sleeves. She constantly saw friends and relatives of clients. A vast amount of personal information can emerge in the give and take of séance conversation, to be fed back to sitters in later séances… Obtaining facts about prominent persons is not difficult. Obituaries can be checked. Courthouses contain birth and marriage records, real estate sales, and so on. Reference books abound in biographical data that sitters often swear a medium could not possibly know.

Even disregarding the fact that Mrs. Piper and her husband were shadowed by detectives to check whether they were in some way researching "clients" (as noted by Gardner himself), there's good reason to doubt this claim based on the original sources. As Frank Podmore pointed out, Richard Hodgson instituted strict controls to lessen the chance that Mrs. Piper could 'get up' information on sitters prior to a séance. For instance, she…

> …did not even know their names… The sittings were fixed sometimes a fortnight, sometimes only two or three days beforehand; the dates were sometimes changed… in one or other of these instances the precautions taken may have been insufficient… But it would be very difficult to suppose that that loophole was always left open, that malign chance favored Mrs. Piper for nine years so punctually that the sittings which have to be written down as failures now number barely 10 per cent…. And it is at least worth remarking that the one series of sittings where it would have

> been least difficult to anticipate the names of the probable sitters and to provide for their advent – the well-known Professors of Harvard, who came when Mrs Piper was under Professor James' direction, – was one of the least successful here recorded.[19]

If, like Gardner, we (rather cavalierly) disregard the confidence of the original researchers that Mrs. Piper was kept in the dark about the identity of the sitters – how then did she collect information about them once she knew who they were? Gardner has a suggestion: "Mediums in a city know one another. Those who patronize one medium usually visit others. At the time there were scores of mediums in Boston, forming a network of scoundrels who passed information freely back and forth." Once again Frank Podmore pre-empted Gardner's 'revelation' by a century, as noted in the *Journal of the SPR* in 1898.

> [I]t was the rule to introduce sitters by assumed names. We had to suppose first, then, that Mrs. Piper was able to ascertain beforehand who were coming, and the exact date of their sittings. That Mrs. Piper should have worked up the dossiers of all the sitters some time before was practically impossible. No doubt it was permissible to assume a freemasonry amongst professional mediums, and that any information obtained by one of the fraternity would be at the disposal of all. But a considerable proportion of Mrs. Piper's sitters were not even numbered amongst the 500 odd Members and Associates of the ABSPR; and *very few had ever been to a professional medium before.*[20] [my emphasis]

Gardner also fails to mention that, as part of their testing, in 1889 the SPR 'removed' Mrs. Piper from the familiar surrounds of Boston and tested her in England for some time, with the same positive results. As Fred Myers pointed out in his report on these experiments:

> Her sitters (almost always introduced under false name) belonged to several quite different social groups, and were frequently unacquainted with each other. Her correspondence was addressed to my care, and I believe that almost every letter which she received was shown to one or other of us. When in London she stayed in lodgings which we selected... We took great pains to avoid giving information in talk; and a more complete security is to be found in the fact that we were ourselves ignorant of many of the facts given as to our friends' relations, etc.[21]

As the reader may be noting by now, the more one is conversant with the original sources, the less impressive Martin Gardner's scholarship and knowledge on this particular topic seems. When he says "books about Mrs. Piper by believers seldom mention her information failures," he is certainly not talking about the original research documents. Take for example William Newbold's report, in which he is at pains to point out that "I have been especially careful to bring into prominence all distinct failures and any other facts which would tend to detract from the surprising character of many of the statements made." It may be worth noting that Gardner himself assiduously avoids mentioning any of Piper's 'information successes' in his own essay...

Studies in Skepticism

If there is one source that Gardner did consult, it is the 1910 book *Studies in Spiritism*, by Professor Amy Tanner and Dr. Stanley Hall. Hall and Tanner had six less-than-impressive sittings with Mrs. Piper, and devoted the bulk of their book to attacking the SPR's investigations into her mediumship. Gardner's reliance on this source is quite obvious as not only does he reference *Studies in Spiritism* on several occasions, but a number of his criticisms are taken directly from the book's pages – despite the SPR researchers having methodically dismantled them some 90 years previous in their reviews of Hall and Tanner's debunking work. For instance, Gardner quotes extensively from *Studies in Spiritism* in claiming that a lot of personal information was given directly to Mrs. Piper through idle talk during the séance:

> Sitters were asked to hold [Mrs. Piper's] hand, its palm close to their mouth, and to speak with a loud voice as if on a long distance phone call.[22] Occasionally the hand would explore a sitter's face or body. Did the hand require shouting because Mrs. Piper was getting deaf? On the contrary her hearing was extremely acute. As Hall reports, she reacted to everything audible – "noise on the street, the rustle of clothing, the sitter's position, and every noise or motion."
>
> By insisting that sitters address the hand in a loud voice, a strong impression was created that Mrs. Piper was "as much out of the game as if she were dead." If the hand could not hear voices in low tones, surely Mrs. Piper could not hear the conversation of sitters. Convinced that the sleeping Mrs. Piper could hear nothing, sitters felt free to talk to one another. Later they would not even remember what they had said. When information from such whispered conversations came out in later séances, or even in the same séance, they would be amazed.

Eleanor Sidgwick directly addressed this criticism in her review of *Studies in Spiritism*, noting that this supposed 'flaw' in the SPR's method seems to have been an outright assumption from Hall and Tanner, based only on their own failings while sitting with Mrs. Piper. In fact, Mrs. Sidgwick was incredulous that Hall and Tanner would...

> ...believe that the records are so incomplete that even important remarks by the sitter are omitted? Certainly at sittings at which I have been present either as manager or sitter this has not been the case. It has never been assumed by me, nor I think by other

> English sitters, that any remark or sound made in the room was inaudible to the trance-personality, and it surprised me that Mr. Hall and Miss Tanner should ever have acted as if this were so.[23]

James Hyslop echoed Mrs. Sidgwick's perplexity at the claim of séance chatter as the source of the 'supernormal' information obtained by Mrs. Piper: "If they supposed that Dr. Hodgson or I went about the experiments whispering and jabbering about them as these authors confess to doing, they are not only laboring under an illusion but might have obtained information to the contrary if they had read the reports carefully. We made it a most careful business not to talk about anything connected with the experiments in Mrs. Piper's hearing either in or out of the séance room...absolutely every word or whisper made during the trance was taken down."[24]

Though the SPR investigators kept stenographic records of the séances whenever possible, Hall and Tanner incorrectly noted that "never in our own or in other Piper sittings was any full record kept of what her interlocutors said. Still less have involuntary exclamations, inflections, stresses, etc., been noted, and even the full and exact form of questions is rarely, if ever, kept". Hall notes in the passage above that full records of his own sittings were not kept, and yet Gardner claims in his article that Hall and Tanner's six sessions were "recorded verbatim". This is wonderfully ironic, given that Gardner bases his misguided 'séance chatter' theory on Hall and Tanner's own misguided claim that no stenographic records were kept by the SPR investigators.

Hyslop's dismissal of Hall and Tanner's 'research' could easily apply to a number of points in Gardner's ham-fisted critique:

> It is the less excusable because the book pretends to show a knowledge of the various volumes published by the Society.... You would suppose from the authors of this book that they had discovered it and that psychic researchers were especially delinquent in this matter. In his first report Dr. Hodgson remarked his habit of making stenographic

> records. (Vol. VIII., pp. 2 and 88.) Professor James had made them before Dr. Hodgson came to this country. (Vol. XIII, p. 2 and American Proceedings, p. 103).[25]

Hyslop then explicitly points out that Richard Hodgson's motivation in taking stenographic records was "with exactly the same desire to enable readers and students to ascertain for themselves whether information had been imparted directly or indirectly by sitters."[26]

Gardner's reliance on *Studies in Spiritism* is unfortunate, as is his misrepresentation of the objectivity of the authors in reassuring readers of his essay that "Tanner and Hall approached Mrs. Piper with open minds". Gardner's statement is rather difficult to reconcile with the introduction to *Studies in Spiritism*, in which Stanley Hall openly states his belief that "Spiritism is the ruck and muck of modern culture, the common enemy of true science and of true religion, and to drain its dismal and miasmatic marshes is the great work of modern culture... When genetic psychology has done its work, all these psychic researches will take their place among the solemn absurdities in the history of thought".[27] Furthermore, Hall deliberately ignored any possible evidential messages during his sittings with Mrs. Piper, noting that he "had no desire whatever to obtain 'test messages', my results from the published sittings having shown their triviality and dreariness and the impossibility of getting down all the remarks and other circumstances which might explain them". Meanwhile, Eleanor Sidgwick paraphrased Professor Tanner's own criticisms in taking her to task for ignoring the 'tremendous influence of a preconceived theory on one's interpretations of the facts': "In fact her presentation of facts and arguments cannot be assumed to be fair without reference to the originals. For...they constantly misrepresent the case and are essentially misleading."[28] In his own response to Hall and Tanner's book, James Hyslop devoted 98 pages to correcting errors, omissions and misstatements of fact.

Unfortunately, Gardner believes that Hall and Tanner's book offers other worthwhile criticisms:

> Hall's greatest scam was presenting [the alleged spirit of] Hodgson, Mrs. Piper's main control at the time, with fictitious information and names. Hall had met Hodgson only once, but he pretended they had been old friends. Hodgson reciprocated by calling Hall "old chap" and by remembering a wealth of events and discussions that never took place. Hall invented a Bessie Beals. Hodgson had no trouble locating her on the "other side."

Both Gardner and Hall seem to place much importance on this 'scam', despite the fact that Hall had asked Mr. Dorr (Mrs. Piper's SPR 'manager' of the time) whether the investigators had tried such a trick previously. Mr. Dorr's answer was that "many have tried foolery and sometimes have succeeded splendidly, and other times have failed. *Controls are very suggestible and very willing to take up any ideas presented by the sitters*, so that they can be very easily taken in"[29] [my emphasis]. Certainly, this suggestibility should give one pause when trusting the words of Mrs. Piper (or any other medium). But given that the original SPR researchers had noted this aspect – it wasn't ignored or never tested, the researchers simply found that it couldn't explain away the convincing 'hits' that Mrs. Piper managed regularly – it doesn't seem quite as great a scam as Gardner would have us believe.

Leonora Piper in 1890

Continuing, Gardner makes much of the fact that "although Mrs. Piper always insisted she never recalled anything that transpired during a seance, Hall noticed a growing coldness in her attitude toward him and his assistant. After this final disastrous scene, she betrayed no hint of knowing how funny it had been." Gardner gives no reference for this assertion, but I assume that he's drawing from one particular passage in Hall's notes from *Studies in Spiritism*: "I fancied that I myself detected a faint left-over trace of coolness toward me in Mrs. Piper, after waking from the third sitting [of six]...like [a] tenuous after-image".[30] Between the 'fancied', 'faint' and 'tenuous after-image', I find it difficult to ascribe too much importance to Gardner's claim of a "growing coldness". Reading Hall's notes, there seems to be no noticeable attitude from Mrs. Piper. Though, if so, given Hall's own obvious attitude (based on the parts of the book he wrote), it seems it would be just as (or more) likely that Hall and Tanner betrayed their personal feelings about Mrs. Piper's mediumship before and after the séance.

If Hall noticed any later 'coldness' from Mrs. Piper, it could well be due to the fact that he and Tanner performed harmful tests on Mrs. Piper in measuring her sensitivity to pain during the fifth séance, though they neglected to tell her afterwards:

> It appears that some time after the [fifth] sitting red spots appeared on Mrs. Piper's palm, and her index finger was numb for two or three days, the red spots being the after-effects of the esthesiometer pressure, and the numbness probably due to the pain-pressure experiments. Her lower lip was also blistered from the camphor used. Her daughter wrote to Mr. Dorr the day after the sitting describing these effects, in some distress, saying that she had hoped that the need for such tests was over...Dr. Hall wrote both to Mr. Dorr and Mrs. Piper explaining the experiments and stating that we had finished them, and that we should not have gone so far had not the Hodgson control authorized us to do so.[31]

The sensitivity experiments were carried out to test Mrs. Piper's trance condition. Gardner himself has little doubt as to the fraudulent nature of this aspect of her mediumship, scolding that "[P]ersons suffering genuine trance seizures do not go in and out of them in theatrical ways calculated to impress audiences." It is worth noting the loaded language here – it is only Gardner's assumption that the seizure was "calculated" to impress audiences, regardless of whether one accepts that it was "theatrical".[32] More importantly though, in doing so Gardner disregards the opinion of the contemporary researchers. For instance, Frank Podmore made clear that "by an almost universal consensus of opinion her trance is a genuine one." But if Gardner feels that the SPR researchers' opinions can't be trusted, he can always turn to his own trusted source – as a result of their tests, Hall and Tanner themselves concluded that "certainly her respiratory functions, taste, smell, general tactile sensibility and motor innervation are asleep".[33] They began by using an esthesiometer on Mrs. Piper's hand: "Again [Professor Hall] pressed, and this time with enough force to give considerable pain if the sensitiveness were normal, but no response came." They then "uncorked a camphor bottle and held it close up to Mrs. Piper's nostrils so that it seemed as if she could hardly get any air unless saturated with camphor. I held it there for some seconds without any effect…I then put about one-third of a teaspoonful of camphor in her mouth…it produced no palpable effect."[34] It certainly produced an after-effect though, as noted above: the camphor blistered Mrs. Piper's lips. Similarly, Richard Hodgson held ammonia to Mrs. Piper's nostrils, but "could not detect the smallest signs of discomfort after [she] had taken several inhalations of strong ammonia." Hodgson nonchalantly added in a footnote that "Mrs. Piper suffered somewhat after the trance was over."[35] (it caused her nose to bleed and brought on chronic sensitivity[36]) The British psychical researcher and well-respected scientist Sir Oliver Lodge wrote in his own account of Mrs. Piper's mediumship that the trance "is, to the best of my belief, a genuine one. In it Mrs. Piper is (sometimes, at least) insensible to pain, as tested by suddenly pushing a needle into her hand, which causes not the slightest flinching."[37] William James noted that her pupils were contracted during the trance state, and Hall and Tanner noted Mrs. Piper's change in breathing "from 20 to 22 for the normal, to between 7 and 10 during the trance, up to the time the hand ceases to write." The famous Nobel Prize-winning French physiologist Charles Richet investigated Mrs. Piper's mediumship and declared her trance to be genuine. Who should we trust more, Richet's personal testimony, or Martin Gardner's speculation a century later?

Gardner's reliance on such a poor source as *Studies in Spiritism* suggests either intentional cherry-picking, or ignorance of the SPR's own extensive, primary sources on this topic. Mrs. Sidgwick, a very sober critic not given to hyperbole, mentioned that "it is likely enough that [*Studies in Spiritism*] will impress those who derive their knowledge of the evidence discussed from it alone; but a very different view will be formed by those who are able to check Dr. Tanner's version of the evidence by reference to the original sources." Those sentiments apply equally to Martin Gardner's essay – it is unfortunate then that Gardner's article will be read by a far wider audience than the original sources, as his negative influence is founded upon what seems to be almost complete ignorance of the case in question.

Real Skepticism

Gardner's criticisms are more successful when he brings up particular instances that concerned the original researchers. He rightly points out that Phinuit seemed unable to provide the contents of letters and secret messages written by the dead communicators he was in contact with: "Three times Phinuit tried vainly to guess the contents of a sealed envelope in James's possession, even though the doctor supposedly contacted the dead woman who wrote the letter." This particular case is a major stumbling block to the idea that the communicators were truly who they claimed to be – surely they would remember their own writings? All the same, there are still interesting facets to the case that are suggestive of some paranormal faculty, which Gardner doesn't mention.[38] But this genuine criticism is an isolated instance in Gardner's essay. If he was more conversant with the source material he certainly could have raised other concerns. One of William Newbold's sittings provided minor evidence of Mrs. Piper possibly purloining phrases from dictionaries. And an odd mistake in a sitting with a Professor Bowditch could be an indication of prior research being done on the sitter. Gardner would have done better to concentrate on analysis of some of these isolated incidents, instead of broad criticisms that have previously been comprehensively rebutted.

If Gardner's essay was the only source consulted on Mrs. Piper's mediumship, the reader would walk away thinking he had revealed her secrets, exposed the original researchers as easily-duped amateurs, and shown that the Piper case is yet another simple case of fraud. Nothing could be further from the truth. While the case is perhaps now too far back in time to ever get to the truth, the SPR's resident curmudgeon Frank Podmore explained in 1899 why it is worth paying close attention to:

> [T]he abundance of the material, the fullness of the records, the watchful supervision exercised over the medium herself for some years past, and the extraordinary and almost uniformly high level of success, make these records much more noteworthy than any previous accounts of the kind... In all these years – now thirteen or more – during which Mrs. Piper has been under the close observation, first of Professor William James, and afterwards of Dr. Hodgson and other competent persons – though she has been shadowed by detectives, though her personal luggage, as Professor Lodge has told us, has been searched, her correspondence read, her goings-out and comings-in closely watched – during all these years not the smallest circumstance has come to light reflecting in any way upon her honesty. Certainly no other medium has been exposed to so stringent an ordeal... Dr. Hodgson himself... [has] succeeded in bringing home the charge of dishonesty to very many professional mediums, that this medium should have passed through the most searching and prolonged inquiries without even a rumour of an exposure, or the discovery of any suspicious circumstances, is a fact entitled to some weight.[39]

The ever-skeptical Podmore goes on to point out that even those cynical of the case should play close attention to it, given it would be a case study of one of the greatest scams of all time, being played out for more than two decades under the close supervision of scientists well-versed in exposing frauds:

> In a word, if Mrs. Piper's trance utterances are entirely founded on knowledge acquired by normal means, Mrs. Piper must be admitted to have inaugurated a new departure in fraud. Nothing to approach this has ever been done before. On the assumption that all so-called clairvoyance is fraudulent, we have seen the utmost which fraud has been able to accomplish in the past, and at its best it falls immeasurably short of Mrs. Piper's achievements. Now, that in itself requires explanation… On the assumption of fraud the tremendous gulf between her and them is an almost insuperable obstacle.[40]

There are other fascinating aspects for discussion in this case even on the assumption of deception. One of the more curious aspects of Mrs. Piper's trance mediumship was that for a time (during the transition from communication via the voice, to communication via writing), three different 'communicators' could hold 'conversations' with three different sitters *at the same time* – one through voice, one writing with the right hand, and one writing with the left hand. Yet Gardner casually explains away this bizarre simultaneous three-way mediumship simply by saying Mrs. Piper was "strongly ambidextrous". And there are other aspects that should give the curious mind pause before dismissing her as a fraud. Given her reputation after the first couple of years of investigation, Mrs. Piper could have left the service of the SPR and charged exorbitant amounts of money offering sittings for the rich and powerful, with much less chance of being caught. Instead, she remained on a compensatory wage under the skeptical eyes of investigators for a good portion of her lifetime. Further to that, if she was a fraudulent medium, why change 'technique' from voice mediumship when it was so successful, to developing simultaneous voice and writing (and at times communicating via mirror writing) for no additional reward or benefit? And how did she fool scientists and physicians that her trance was genuine, showing no reaction to pain sensitivity tests including surprise needle jabs, flames held to her skin, and long inhalations of ammonia?

Martin Gardner could have engaged readers with any of these topics, but instead he steers them away from intelligent discussion of this case. He ignores the source material completely, raising criticisms that were comprehensively dealt with a century before he wrote his essay, and impugns the integrity of the original investigators despite having no grounds to do so. If Gardner is, in the words of Stephen Jay Gould, "the single brightest beacon defending rationality and good science against...mysticism", one would have to think that rationality and good science are in serious trouble on the evidence offered in this particular essay. By Marcello Truzzi's definition, given at the start of this article, Gardner here is certainly being a 'pseudo-skeptic'.

Probably the most egregious problem with Gardner's essay that I have mentioned – amongst many – is his convenient avoidance of Richard Hodgson's role in the investigation of Mrs. Piper. Fittingly, Hodgson's introduction to his first report on her mediumship addresses (and rebuts) the bulk of Gardner's revelations about Mrs. Piper's 'fraudulent techniques' in one simple sentence. Hodgson stated that in undertaking his investigation, he was...

> ...compelled to assume, in the first instance, that Mrs. Piper was fraudulent and obtained her information previously by ordinary means, such as inquiries by confederates, etc...further information given in various ways by the sitter, consciously or unconsciously, by speech, gesture, and other muscular action.[41]

My critique of Gardner's essay examines just one case covered by Martin Gardner, amongst a life-time of writings on a variety of topics, and so it would be wrong to extrapolate too far from it, or say that it detracts in any major way from his positive contributions to academia. This topic is the only one that I personally know the source material well enough to feel comfortable rebuking Gardner's attempt to gloss over years of scientific investigation by the SPR researchers. What I hope it does show is that the claims of prominent skeptics should be subjected to as much scrutiny as any other person; there needs to be a change to the blind acceptance of the words of individuals who claim ownership of the title "skeptic". This might best begin within skeptical organizations themselves, by promoting more internal criticism and an end to the demagoguery that pervades the likes of CSI(COP) and the James Randi Educational Foundation (JREF).

Replies From Beyond the Grave

In "How Mrs. Piper Bamboozled William James", Martin Gardner tells how the famous academic first discovered Mrs. Piper:

> In 1885 they had an old Irish servant whose sister worked for a prominent Beacon Hill family often visited by James's mother-in-law. Yet when the mother-in-law sat with Mrs. Piper, William was flabbergasted to learn that her controls had named members of his family!

Let us compare Gardner's summation of William James as being "flabbergasted" (with exclamation point no less), with how James actually reported this moment, as noted in Robert Richardson's biography *William James: In the Maelstrom of American Modernism*: "Hearing these things from his sister-in-law and his mother-in-law, James's response was skeptical. 'I remember playing the *esprit fort* on that occasion before my feminine relatives and seeking to explain by simple considerations the marvelous character of the facts which they brought back.'"

Funnily enough, the only thing that James was "flabbergasted" about was how *little* the medium seemed to be able to produce about his famous family (beyond his own fame, his brother was Henry James, the eminent 19th century novelist):

> The sceptical theory of her successes is that she keeps a sort of detective bureau open upon the world at large, so that whoever may call is pretty sure to find her prepared with facts about his life. Few things could have been easier, in Boston, than for Mrs. Piper to collect facts about my own father's family for use in my sittings with her. But although my father, my mother, and a deceased brother were repeatedly announced as present, nothing but their bare names ever came out...

Given the title and focus of Gardner's article, it is perhaps worth giving William James a right of reply, allowing him to defend himself from beyond the grave. Gardner, through omission and misrepresentation, portrays James and the other investigators as inept and falling over themselves to accept the afterlife theory, when this was far from the truth. It seems that self-appointed skeptics never change, as Columbia University academic James McKeen Cattell (then President of the American Psychological Association) claimed much the same things as Gardner at the time (with much the same 'evidence'), and William James's response could just as easily be applied to Gardner's essay. James responded to Cattell (and in may ways, Gardner too) by saying:

William James

> Your reference to my name...justifies me in making some remarks of my own... Any hearing for such phenomena is so hard to get from scientific readers that one who believes them worthy of

careful study is in duty bound to resent such contemptuous public notice of them in high quarters as would still further encourage the fashion of their neglect.

...The scientific mind is by the pressure of professional opinion painfully drilled to fairness and logic in discussing orthodox phenomena. But in such mere matters of superstition as a medium's trances it feels so confident of impunity and indulgence whatever it may say, provided it be only contemptuous enough, that it fairly revels in the untrained barbarians' arsenal of logical weapons, including all the various sophisms enumerated in the books.

Your own comments seem to me an excellent illustration of this fact.

...what name should we assign to the fallacy by which you quote one of those five sitters as saying that he himself got nothing from the medium 'but a few preposterous compliments,' whilst you leave unquoted the larger part of his report, relating the inexplicable knowledge which the medium showed of the family affairs of his wife, who accompanied him to the sitting? I am not sure that the logic books contain any technical name for the fallacy here, but in legal language it is sometimes called *suppressio veri*, sometimes something still less polite. At any rate, you will admit on reflection that to use the conclusion of that sitter's report alone, as you did, was to influence your reader's mind in an unfair way.

...Please observe that I am saying nothing of the merits of the case, but only of the merits of your forms of controversy which, alas, are typical. The case surely deserves opposition more powerful from the logical point of view than your remarks; and I beg such readers... as care to form a reasonable opinion to seek the materials for it in the *Proceedings of the Society for Psychical Research*, Part

> XXXIII (where they will find a candid report based on 500 sittings since the last report was made), rather than in the five little negative instances which you so triumphantly cull out and quote.[42]

Unscientific skepticism of the type exhibited by Gardner and Cattel is a corrosive one which, rather than defending science, instead shields it from possible new discoveries and viewpoints through irrational over-protectiveness. It also brings skepticism as a whole into disrepute when such cheap tactics are employed. In his article "How Mrs. Piper Bamboozled William James", Martin Gardner ignores the original scientific work done, misrepresents the competency of the investigators, and misleads the reader both through incorrect statements and loaded language. This is hardly the type of writing we would expect from "one of the great intellects produced in this country in this century."

Sadly for Martin Gardner, perhaps the most succinct summary of his essay can be found in James Hyslop's caustic response to Hall and Tanner's *Studies in Spiritism*, written nearly 100 years previous:

> The calm critic can only say that the book either displays the grossest ignorance of the facts and the subject, or it is a colossal piece of constructive lying. The authors may take either horn of the dilemma they like.[43]

Greg Taylor is the owner and editor of the online alternative news portal, *The Daily Grail* (www.dailygrail.com), and is also the editor of *Darklore*. He is widely read in topics that challenge the orthodox worldview, from alternative history to the mysteries of human consciousness. Greg currently resides in Brisbane, Australia. His first book, *The Guide to Dan Brown's The Lost Symbol*, is a guidebook to the esoteric history and locations featured in the best-selling sequel to *The Da Vinci Code*.

The Fairies' Portal

Close Encounters of the Little Kind

by *Theo Paijmans*

The most mysterious aspect of the UFO phenomenon is found in the accounts of close encounters with their occupants. It is one thing to suddenly interact one evening somewhere on a deserted stretch of road with a luminous something that has just descended out of the dark night. It is another to actually meet with that which comes forth from it or seems connected to it. An entity that draws you into a reenactment of a primal play staged countless times since time immemorial. The experience may be so profoundly alien, if you will excuse the terminology, that in searching for analogies to express what we have just experienced, we may find ourselves drawn to ancient descriptions of angels and demons in the musty and forbidden tomes of yore.

Reports of 'alien' occupants have been a recurring feature of UFO sightings in the 63 years since Kenneth Arnold's famous account, but there are also many tales of encounters with remarkable creatures and humanoids which predate this milestone. These mentions are not restricted to later published recollections, but can also be found in abundance in contemporary newspapers. The study of anomalous lights and objects in the skies began in earnest with Charles Fort's seminal *Book of the Damned*, first published in 1919. But we can also find intriguing accounts of encounters with denizens of other planes of existence in the vast body of recorded ghostlore and in other improbable tales recorded before and after the publication of Fort's work.

This paper examines some of the accounts that were reported in 19th and early 20th century newspapers. While originally planning to cover a wide variety of 'entities' of various classifications, due to space limitations I have limited the scope of this article to the treatment of accounts of 'little men' and dwarfs. I have deemed it necessary to report these encounters as objectively as possible, rather than accepting the original, subjective explanations of these manifestations as being ghosts, phantoms and apparitions. This was an explanation that obviously befitted the times in which these incidents were alleged to have occurred. My objective is to shed these tales of their early interpretations, and look at them in the relatively fresh light of the Close Encounter phenomenon. As a result we find ourselves challenged to formulate new interpretations. Moreover, some of these early accounts provide context for a number of the most puzzling later reports. Accounts so singular in nature that, until now, they stood in isolation, lacking any contextual framework of comparative events. This new development gives rise to the thought that some recollections – improbable as they may seem – must either harken back to a now long lost folkloric tradition, or are actual, truthful tales of some very unusual events, told in all honesty and as best as the persons could remember them.

Litte Green Men and Little Red Devils

What is marked as "one of the most bizarre not to mention unpleasant humanoid reports of all time" occurred at Farmersville, Texas in May 1913, though it first came to attention in 1985 via the grandson of the man who had experienced it as a child. The old man was later interviewed about the incident, and while the interviewer could not accept the bizarre tale, he had no doubt in the sincerity and sanity of the witness. The experience occurred when he was 12 years old, living with his two brothers and the rest of the family two miles outside Farmersville on a cotton farm. One morning they were out chopping cotton, when their two dogs began barking furiously some 50 to 75 feet away on the other side of a picket fence. The brothers went to the dogs to investigate and were shocked to find a little man who...

> ...looked like he was sitting on something. He was looking toward the north. He was no more than 18 inches high and kind of a dark green color. He was the same smooth color all over. He didn't seem to have any shoes on, but I don't really remember his feet. His arms were hanging down just beside him. He had on a hat that reminded me of a Mexican hat. It was a little round hat that looked like it was built onto him. He didn't have on any clothes. Everything looked like a rubber suit including the hat... He just stood still. I guess he was just scared to death... Right after we got there, the dogs jumped him.

The dogs tore the little man apart, spilling what looked like human organs. The next day all traces of the strange little man were gone.[1]

Eight years previously, the city of Philadelphia was in uproar over the disappearance of 19-year-old Harriet Pauline Bald. She had been missing since March that year after complaining of a pain in the back of her head. She returned home after an absence of sixteen months, in August, 1906. The reason for her sudden disappearance, she stated in the *Philadelphia Inquirer*, was a little green man:

> Prior to my leaving the city I was ill. I was very nervous and sleep was utterly impossible. Lying awake in bed one night, shortly before I crept out of the city, I was suddenly startled by what seemed to me to be a little man, dressed in green clothes and with red hair, jump in the window. After gazing about the odd form hopped to the mantelpiece. Grinning, the unearthly creature pointed his finger to me and said: 'Harriet! Go away and hide yourself!' That all was surely a dream, but so firmly did it impress itself upon my mind that for hours I could think of nothing else. I guess that I might attribute my home-leaving to the mythical demon. Just to think, to see that unsightly creature sitting upon the mantelpiece, by the foot of the bed, and beckoning and urging and pleading with me to go away! It may sound funny, but that red-haired, green-garbed devil, I firmly believe, was the cause of my going away so abruptly.[2]

In the last week of September, 1907, something described as a "little red devil" badly frightened the children of the Sacred Heart school in Dubuque, Iowa:

> A mystery of peculiar character has developed in the Sacred Heart school on Eagle Point avenue and while the priests in charge of the parish and the sisters at the school do not feel particularly alarmed about it, there is something that has caused them to take precautions to prevent the children from being frightened again, as they have been for the past two nights. What the thing is the teachers do not know. Several of the sisters are inclined to lay the whole thing to the imagination of the children. One thing is certain, however, and that is that several little girls, who have remained to dust the school room after school hours, have been greatly frightened and one little tot fainted when she saw the sight.

> After school on the Wednesday afternoon (September 25), several of the little girls had remained after hours to help dust the rooms and put the school in order for the following day. While they were at work one of them had her attention attracted to the second floor of the building where she saw what she is positive was the form of a person dressed in red and sitting at the head of the stairs. The little one was so frightened that she fainted and had to be taken into the school by older persons who were about the building.
>
> Thinking that the child was rather nervous and that she was the victim of a hallucination, the sisters in the school had several other girls remain Thursday evening. About 5:30 one of the girls had her attention attracted to the same spot and sure enough there was that figure sitting in the window. The child ran screaming to the school house and told the other girls of what she had seen. They all left the building at once. A mother of one of the girls states that her child came home in nervous excitement and stated that she had seen a "little red devil sitting in the window of the school."
>
> In view of the fact that different children, on different evenings, had seen the same object, the authorities of the school decided to place a man on guard at the place Friday evening.[3]

The sighting of a red devil reminds one of the tales involving another red imp; that of the *Nain Rouge*, or Red Dwarf of Detroit. Said to always appear at times of distress and calamity, a 1923 newspaper described the creature thus: "The Red Dwarf had a red face and protruding teeth", and had been seen "as late as 1905 when men and women in terror at the fire that consumed the city saw his squat shape moving thru the smoke grinning."[4]

Closer to the 1913 Farmersville incident, we find an account of a "weird apparition" that plagued the residents of Bridgeboro, New Jersey in 1911: "Joe Johnson and a party of friends, passing what is known as the Paxson pits a few nights ago, were confronted by a dwarf, which, they declared, climbed out of a pit. For a minute it lingered and then, giving a piercing scream, vanished into the woods. Other witnesses, farmers of unquestionable veracity, substantiate Johnson's story." The explanation on offer was that there might be a connection with an accident 'years ago' where a young boy broke his neck when the carriage he was in overturned and he fell into the pit.[5]

The Gnome and the Moving Light

As the accounts above illustrate, we are not left empty-handed in establishing a possible context for the 1913 Farmersville incident. There are several puzzling similarities. Children or near-adults often feature in these accounts, and I found numerous accounts even predating these early 20th century ones. Take, for example, this mention of a 'ghost' that measured three feet in height that infested the environs of North Haven, Connecticut in January 1886, and which reads very similarly to the Farmersville incident:

> North Haven, a quiet village...is excited over the appearance of a ghost. It is described as 3 feet in height, and wears a velvet suit and hat. Oliver McNulty, a laborer in Shears's brickyard, with four of his companions, ran across the ghost a night or two ago, when they followed up a moving light. The next day the men happened to be near the same spot and again discovered the gnome. McNulty struck it with his spade. He declares that he cut it in two, and that the severed parts reunited and then disappeared alltogether. Old residents of the town say that they have seen it at different times for 30 years. McNulty's story is indorsed by the four men who were with him. The neighborhood is low, level, and damp, and the *ignis fatuus* is frequently seen there.[6]

Those who have read my *Darklore 2* article, "The Cohorts of the Damned", may also remember the strange tale of the abduction of the daughter of Andy Crowe by three 'fairies' dressed in black, who drove her around in an invisible carriage. Published in 1886, the incident is said to have occurred at Dubuque, where in 1907 the little red devil was seen by several children.

To return to encounters with dwarflike entities, we have an account that same year of a strange encounter on a deserted road at night near Millington, Connecticut. It was around seven or eight at evening and Millington resident Edgar Martin was travelling home, when suddenly his oxen...

> ...jumped to one side... Mr. Martin, too, was startled out of his senses, and looking about him to discover the cause that had excited his cattle, saw, apparently walking some twenty feet ahead of his leaders, a little dwarf-like man not more than 3 feet tall, as he appeared in the dim light. The little object as he looked from behind, seemed made up principally of a long flowing cloak which hung from his shoulders and spread out at the bottom until its wearer looked like an animated "A", on whose head was perched a little jaunty peaked hat, and in whose hand was a long cane perhaps double his height in length, which seemed to assist his little duck-like, long-stockinged legs, which were visible as the wind raised and lowered his flowing cloak. Diminuitive as the peculiar little figure was, he seemed to have no trouble in keeping the same distance ahead of the team that he held when first seen...'

Martin jumped from his cart, ran towards the figure who did not respond to his hailings, and struck it with his whip. The dwarf disappeared in an instant, but travelling home further, Martin felt as if "invisible things were about him, in front of him, behind, beside him." His ordeal lasted till he finally reached home.[7]

Accounts such as the ones cited above have their correlates within the later close encounter reports of ufology. Dwarf-like entities and, in particular, 'little green men' are potent archetypes in the lore of ufology. Perhaps this explains why we find such reports mentioned and published in specialist ufological publications, despite there being – on the surface at least – no overt connection to any manifestation of a UFO itself. There is no mention of a nearby brightly lit, otherwordly flying object that dazzled the senses, usually considered an essential trapping of a 'Close Encounter of the Third Kind'. Yet there is an urgent sense of intuition, of instinct, that has prompted many a ufologist to include this type or report, many times more than we are able to list here.

For instance, 'Witness Zero' of the modern UFO phenomenon, Kenneth Arnold, gave this account at the International UFO Congress in Chicago in 1977:

> An Ellen Jonnerson over in Canby, Oregon, a University of Oregon graduate and a very intelligent person, actually saw a little man. She was stunned to see this little man walk across their breezeway and, standing straight up, pass underneath the running board of a '37 Dodge car. He couldn't have been more than nine inches tall and he had features like a man. He wasn't threatening. He was very dark-skinned; he had on a type of little romper and a sort of plaid shirt. She got a good look at him and when I asked her if she didn't feel like she'd like to get closer to him, she said, "I think I should have run over and grabbed the little fellow." But she walked out into the grass and she never saw him again.[8]

However, only rarely in these early accounts do we catch a glimpse of simultaneously occurring anomalous phenomena, such as an entity sighting along with a description of strange lights. One example is the *ignis fatuus* (a term referring to phosphorescent lights sometimes seen hovering over swampy ground – quite literally, the phenomenon J. Allen Hynek would later refer to as "swamp gas") mentioned earlier in the case of the 1886 North Haven Ghost. Another, more surreal account discusses an incident that allegedly occurred in October 1954 in Los Angeles, in which a man reported seeing a flying saucer land in MacArthur Park and…

> …a little man in a white suit get out. A truck then came and carted away both the saucer and the little man, the observer said. He would have gotten the truck's licence number, he said, except that the fog closed in and he couldn't see anymore. The witness refused to give his name "because everybody would think I was crazy."[9]

But before discussing this topic further, a quick digression…

Leprechauns like the wind

Sightings of 'little men' are certainly not restricted to the Americas during this period. This is a worldwide phenomenon, with dwarf-like beings mentioned under various names in the local folklore of many regions. For instance, in April 1908, northwest Meath in the district of Killough in Ireland was in uproar as reports had it that a little man dressed in red had been seen. According to a contemporary account in the *Washington Post*, "Farmers profess to have seen a Leprechaun playing about the neighborhood frequently in the last few days."[10] Three children returning from Sunday school were "suddenly confronted by what they described as a little man wearing shoes, gray stockings, corduroy knickers, a red coat and a green cap." One of the children tried to grasp him but the Leprechaun eluded him, disappearing with a "quaint laugh". No trace of him could be found.[11] The mystery was finally solved when the local police apparently managed to capture the little man two months later. "This modern specimen of the ancient little people of Ireland proves to be a dwarfish man, who eats greedily, but does not talk, answering questions put to him with grunts and squeals."[12] Thirty years later West Limerick found itself the center of attention by the little people. "Crowds, it was stated, were assembling in the evenings at cross-roads hoping to catch a glimpse of the 'good little people'; boys and men had chased the fairies…while a youth named Keely said he had actually held a leprechaun by the hand." As the story had it, the boy, John Keely, saw a solitary little man, who replied to Keely's questions saying "he was from the mountains and it's all equal to you what my business is". The next day, two 'leprechauns' appeared at the crossroads between Balingarry and Kilfinney. Carrying skipping ropes, "they could leap the height of a man", according to eyewitnesses Robert and John Mulligan and others. The little entities allowed Keely to approach them and he took one of them by the hand and set off along the road with him, but when the leprechauns spotted the others lying in wait in the nearby bushes they took fright and "away they went like the wind', with the Mulqueens, Keely, and others in hot pursuit".[13]

From France comes the tale of an incredible encounter in the final year of World War II. Although it lacks all the necessary elements to truly rank it among the ufological close encounters, it nevertheless was reported to a prestigious French UFO research organisation and subsequently published in a French specialist ufological publication. As the story goes, a French priest – only identified as "M. l'abbé X" – was searching for mushrooms in the countryside near Renève, Côte d'Or, Burgundy, France, on the afternoon of April 20, 1945, when he suddenly stumbled upon a little man a mere six inches tall. According to his recollection, written down many years after the event, it carried a little spear or staff and except for its length was otherwise normal in appearance. The 'dwarf' walked as if in a hurry and was only one foot away from the priest. The little man seemed about 70 to 75 years old and appeared frightened and out of breath. It had gray hair and a small beard and wore dark-red clothes covering his body, hands and feet, though his head was bare. His staff extended about an inch above his head. The priest considered capturing the little man, but he suffered a reluctance much like Ellen Jonnerson in Canby experienced during her encounter. His rationale was that he did not want to risk being wounded by the little man's staff. After about 20 or 30 seconds the dwarflike entity disappeared into the nearby thickets before the priest had an opportunity to communicate with him.

The cleric wrote a letter describing his encounter to G.E.P.A., the 'Groupement d'Etude de Phénomènes Aériens,' in 1975. G.E.P.A. sent a member near Renève to interview the priest, who also went to Paris to tell of his encounter. The cleric explained that as a Catholic priest the belief in the reality of apparitions was part of his mental upbringing, but he was certain that he had not experienced such, and instead had encountered a solid entity. The priest also said that, had he placed it in the context of a religious vision, it would have amounted to sacrilege to capture the little man.[14]

The winged ship

In nearly all of the accounts that we have mentioned thus far we have noted a suspicious absence of the UFO itself. Is it then possible to draw any correlation between these early reports of little men of various shapes, sizes and clothing, and the UFO phenomenon? In the 'proto-UFO' timeframes (that is, before the Arnold sighting) that we have covered there *were* sporadic outbreaks of anomalous aerial phenomena that did include what we may consider to be true Close Encounters of the Third Kind. For instance, in August 1906 John Warner, an old soldier and resident of the suburb of Orinoco, insisted he saw an airship. It was cigar-shaped, painted green and carrying green lights. It landed in his garden, and in it, so Warner claimed, were four men who stated that they were on their way to New York, having come from Chicago. After having asked him for the right direction, "they turned on their power and sailed away."[15]

A year later, in 1907, one Walter Stephenson had a close encounter at the Dikeman Springs in Tennessee. He was…

> …just finishing a long chase with his dogs and sat down on a log to rest, when he espied upon the eastern horizon a speck, which he took to be a large kite. He paid little attention to the object, and shifted his gaze temporarily to other scenes. Soon his attention was attracted to a whirring noise, and looking upward, he saw that the speck which he had a few moments before discovered in the eastern sky had approached almost directly over him, and that the object was in reality a huge balloon, but of a pattern and appearance he had never in his life before seen. He discovered that the floating mass was rapidly approaching the earth. Of a sudden, the observer says, strains of music calculated to charm the spheres burst from the balloon, which circled round and round and finally landed at Kideman springs [sic]. A number of strange people emerged from the car, which was closely curtained with a substance that fairly glistened in the sunshine

> that temporarily burst through the obscuring clouds, and all going to the big, flowing spring, knelt by it in a supplicating attitude and so remained for a minute or more. Mr. Stephenson says that while this was going on he sat quietly within speaking distance, and when the strange visitors arose to their feet and he supposed their devotional excercises were over, he asked if he might be permitted to inquire who they were, and what their mission? He said that instantly a visard was lifted by one of the company and the benign face of a lady showed from underneath and said in German: "Haben sie Beten?" (did you pray?") and instantly all were aboard, the airship rose, circled about for a minute or more, and was gone in a westerly direction.

According to Mr. Stephenson, "the incident left an impression upon him that he can never forget, and while he knows that it was some human invention, it looked and the music sounded more like that of angels than of mortals."[16]

In June and July of 1908 Denmark was also visited by phantom airships,[17] and just a year later during a flap over England and Ireland we find mention of two close encounters. But it is one thing to find references to sightings of unusual aerial phenomena and contact with their largely 'human-like' occupants, and another to find early reports of weird encounters with implausible *little* people that also include some sort of UFO sighting. On the surface, there seems to actually be no connection. Yet long before Jacques Vallee covered the topic in his seminal 1969 book *Passport to Magonia* – with its little-mentioned subtitle *From Folklore To Flying Saucers* – this unusual theme drew attention from a surprising source: it is in fiction that we discover direct links between 'the little people' and advanced aerial constructions. One of the earliest examples can be found in a book that was published in the famous 'flap year' of 1897, so it seems likely that the writer was inspired by these events. The book is titled *The Winged Ship*, and in it, the brownies build an airship and travel with it through the ether. There are more contemporary examples;

From Solon L. Goode, *The Winged Ship*, illustrated by Vinnie Ream Goode

He had slipped in the pantry that morn on the sly
And had eaten the whole of a big pumpkin pie.

The soldiers were dazed with the shower of white.

Illustrations from *The Winged Ship*

in September 1908 an American newspaper published the children's tale "Aggie and the Tree Fairy", replete with an illustration of the little girl taken by the hand by a female fairy with wand, "sailing through the air toward a beautiful airship".[18]

Children of the time were inspired by this theme too. As was their way, the old newspapers often published stories written by children on their youngster pages, or in connection with a contest. In 1910 Ethel Holsinger saw her story "The Moon Fairies' Visit to Mars" published in the *Washington Post.* In her narrative, the Moon Fairies travel to Mars by way of a cloud; they meet the Sun fairies, the cloud fairies "and the good fairies of Mars".[19] In 1925, 12-year-old Ruth Hanson won first prize with her short story "On The Green Cheese". Her story opens as follows:

> Swinging idle in the hammock one June day, I sat up with a start, for there, perched on the edge of the hammock was a tiny elf, who was poking with his cane [*sic*]. He wore a red coat trimmed with pink oak leaves. His trousers were of a sky-blue cloth, spun by Mrs. Bug, out of her silken web. Under his fawn-colored cap peeped his tiny face, which reminded me very much of a dried apple. Beside him was his airship, fashioned from a green leaf with small twigs protruding from the sides supporting the silken wings woven from a spider web. "How would you like to go on a trip to the moon this evening, with me?"asked the elf.[20]

These fictional stories also appeared as bedtime stories for children in the newspapers. In 1914 a short tale was published under the title "The Fairies' Trip to The Stars",[21] and a year later we find the following scene in the tale "On The yankee Doodle". A boy has built a make-believe airship from some kitchen utensils, when suddenly…

> …From somewhere came a voice the like of which they had never heard before. "Hello, big children!"it said. "Oooo!" exclaimed

> Charles, "look up there!" And he pointed to a branch overhead. Then they saw a tiny man sitting astride a twig. He was dressed in blue with gold braid on his sleeves. He was not more than four inches tall.

Asking the children if they would like to visit the places they dreamt about in their make-believe airship, the fairy – for that is what he is – climbs down, looks at his watch, waves his little stick, and off they fly.[22]

As these examples from children's literature show, we are enchanted from an early age by the possibility of supernatural visitors, and also the idea of flying in strange contraptions to faraway places. That might explain the prominence of children in these accounts of meetings with little men (though why are they almost never little women?). The rich fairy lore of various cultures allows us to temporarily transcend 'mundane reality', and keeps us aware of other realms beside that which mankind inhabits. So it is not illogical to assume that somewhere along the way, with the advent of the proto-UFO phenomenon, these two became intertwined. This entangling of the

Illustration from "Aggie and the Tree Fairy" (*The Edwardsville Intelligencer*, 1908)

ingrained mythology of fairies and little people – or more precisely, the 19th century Victorian image of those mysterious beings (which found full voice in the 'Cottingley Fairy' photos) – and the nascent UFO folklore might explain why we find such an incident as the one at Farmersville in 1913 mentioned in a UFO encyclopedia, and why it was first reported by an American UFO organisation. Or why a deeply puzzled Catholic priest finally broke his silence and contacted a French UFO research and investigation organisation about his 'fairy' sighting. After all, in these accounts, there is no mention of a UFO. These tales have a profound resonance, and as we have seen by their literary precursors, they answer a deeply felt 'otherworldly' yearning. To be transported in a magnificent airship by beautiful fairies to other worlds is a siren's call that few can resist.

Theo Paijmans has authored two books, a history of the UFO phenomenon published in 1996, and *Free Energy Pioneer: John Worrell Keely*. Published in 1998 it has been reprinted and saw a translated version published in Japan in 2000. His articles and papers have appeared in various publications including *All Hallows, Strange Attractor, The Anomalist, Gazette Fortéenne, The Centre for Fortean Zoology Yearbook 2009* and *Fortean Times*, where he is a regular columnist. His papers on the history of Spring-heeled Jack in America and Spring-heeled Jack in New Zealand will appear in an academic collection under the editorship of Mike Dash. Theo was invited to speak and participate in the panel discussion at the 2008 Fortean Times Unconvention, and has recently appeared as expert in a Discovery Channel documentary on the Vril Society (*Dark Fellowships: The Vril Society*), the subject of his book slated to appear at a later date.

"Under Clouds" (Image courtesy Alex Andreev www.alexandreev.com)

ENCHANTED ISLANDS and SHIPS IN THE SKY

A critical look at UFOs in folklore

by *Nigel Watson*

The idea that extraterrestrials visit us in spaceships is a relatively new one. However, myths, legends, history and folklore indicate that humanity has been under constant surveillance by 'others' since the dawn of civilization – or at the very least, that 'they' come to visit us now and again to check up on our evolutionary progress. One such line of thinking suggests that the surge in UFO sightings since the 1940s is due to our development of nuclear weapons that could eliminate all life on this planet (and, perhaps, others).

Vimana flying craft are described in the ancient *Samarangana-Sartradhara*, *Mahababaharata* and *Ramayana* Indian texts. The various propulsion systems and means of controlling the vimana are detailed in Chapter 31 of the former text, and the other two epic poems indicate that they travel on rays enabling them to carry death anywhere in the world.[1]

The Greeks had whole hierarchies of deities whom, although superhuman had very human forms of social organization and behaviour. Such deities were thought by ancient cultures to live on mountaintops, in underground caverns, or on enchanted islands and other locations that are generally inaccessible to humans. From a ufological perspective, it is easy to regard the golden chariot of Hades as a flying saucer that he used to abduct people to his underground base. In this sense, the new deities are aliens who lurk in hellish underground bases and the hollow earth, and/or they are heavenly creatures from Eden-like planets.

The Bible is said to contain many accounts that refer to meetings with Space People and thereby helped create the religious rules and beliefs that have shaped Western civilization. In Chapter 2 of the *Second Book of Kings*, it has the story of Elijah who saw a fiery chariot that tore asunder. This was followed by a great whirlwind that sent Elijah skywards to heaven. It is easy to see why he is often cited as one of the first abductees. In the *Book of Ezekiel*, UFO-type objects are reported on four occasions. The most striking sighting by Ezekiel was of great cloud issuing forth flashes of fire accompanied by stormy winds. Inside this, he saw four wheels with eyes, and four sparkling bronze, winged men. Above them was a throne with a man upon it.

Enoch was visited by two tall, strange looking men who gave secret knowledge about God and took him on a tour of the seven tiers of heaven. Ufologists note that the *Second Book of Enoch* also said that Enoch thought a few days had passed but when he returned many centuries had passed. This seems to indicate that he really visited other planetary realms rather than tiers of heaven. Although expunged from the Bible by St. Augustine, the *Book of Enoch* refers to several visitations of aliens, or as they saw them angels from heaven. It tells of how the angels lusted after the comely daughters of humankind

and took them as their wives. The results of this union were giants who turned against humanity.

The corruption and wars of our world caused this injunction from the heavenly creatures, "...from henceforth you shall not ascend into heaven unto all eternity, and that judgment has been finally passed upon you."[2]

Contactees and abductees often cite the *Book of Revelations* with its apocalyptic visions of doom, and many writers in the 1950s and 1960s noted that the Bible and other ancient texts could have been reporting alien encounters and UFOs rather than religious visions.

These ideas gripped the public imagination with the ancient astronaut theories of Erich von Daniken outlined in his best-selling book *Chariots of the Gods?* At that time, in the late 1960s, the Apollo astronauts were preparing for our own moon landing, which made his reinterpretation of the Bible perfect sense at the height of the secular Space Race.[3]

Classical Greek and Roman texts mention "burning shields" and other wonders in the sky. These visions in the sky were often reported as gods, angels, serpents, dragons, signs and wonders. Over the passage of time, it is even harder to determine if they are misinterpretations of meteors, comets, and meteorological phenomenon or simply fiction for the purpose of entertainment or political satire.[4]

Return from Magonia

In the Middle Ages, a country or region beyond the clouds called Magonia was believed to be the home of flying ships seen in our skies. Since the publication of Jacques Vallee's influential book *Passport to*

Magonia, ufologists have used the term Magonia to highlight the historical and cultural context of UFO sightings.[5] This was reinforced by the adoption of *Magonia* as the title of the British magazine that was originally called the *Merseyside UFO Bulletin* (MUFOB). Since then, the notion of Magonia has spread its influence to our present-day concept of alien abduction and the nature of the aliens themselves.

One of the most influential accounts was given by Agobard (779-840 A.D.), Archbishop of Lyons, who reported the capture of four people who had fallen from 'cloud ships.' He reported:

> We have seen and heard many men plunged in such great stupidity, sunk in such depths of folly, as to believe that there is a certain region [some accounts say country], which they call Magonia, whence ships sail in the clouds, in order to carry back to that region those fruits of the earth which are destroyed by hail and tempests; sailors paying rewards to the storm wizards (tempestarii), and themselves receiving corn and other produce.
>
> Out of the number of those whose blind folly was deep enough to allow them to believe these - three men and a woman who said, they had fallen from these same ships; after keeping them for some days in captivity they had brought them before the assembled multitude, as we have said in our presence to be stoned. But truth prevailed.[6]

Why the strangers claimed to come from the sky and risk being stoned to death is impossible to discern, but at least Agobard, like a modern-day sceptic, was able to keep the crowd's impulses in check.

Most of the UFO literature ignores the fact that the full Latin title of Agobard's book was *Contra Insulam Vulgi Opinionem de Grandine et Tonitruis* ("Against the Absurd Notions of the Common People About Hail and Thunder"), and they do not mention the sentence that precedes the quote above:

> In these districts, almost all persons, noble and plebeian, townsmen and rustics, old and young, believe that hail and thunder may be produced at the will of man - that is, by the incantations of certain men called tempestarii...

It was thought the crops destroyed by these tempestarii were used to trade with the aerial sailors of Magonia. Agobard's intention was to denounce such ideas as blasphemous because only God had the power to control the elements, and he did not like the idea that credulous people paid counter-sorcerers to prevent hailstorms. He was against pagan beliefs, Jews and anything that undermined Christianity.

Jean-Louis Brodu in his comprehensive study of Magonia deflated several other assumptions about this land of cloud ships.[7] One is that Magonia was known everywhere in the Middle Ages as a land of magicians, whereas it was a term only used in the region of Lyons.

He showed that Agobard's text has been used and distorted for different purposes long before ufologists got hold of it. *Le Comte de Gabalis ou entretien sur les sciences secrets* ("The Count of Gabalis, or Briefing on the Secret Sciences"), written by the abbot Montfaucon de Villars in 1670, used it to confirm the existence of elementary beings of the air called Sylphs. He wrote:

> The Sylphs seeing the People, the Pedants and even the crowned Heads taking side against them, decided to change the bad opinion people had of their innocent fleet by abducting men from all places, to show them their beautiful women, their republic and their government, and to put them back on Earth at different spots of the World.

> When the abducted people were returned by the Sylphs to Earth: The people who saw these men coming down would run towards them from all directions and were so prejudiced as to think that they were Sorcerers dispatched from their fellows to throw venom

> on the fruits and into the springs; in the furore brought about by such imaginings, they took these innocents to be tortured. They thus killed an incredible number of men by water and fire throughout the whole Kingdom.

Villars then used the account by Agobard to support these statements, and to allege that the stories of Fairies in the following centuries were really encounters with Sylphs and Nymphs. In the light of modern-day abduction studies, this fleet of aerial ships abducting men to meet "beautiful women" in their magical realm sounds all too familiar. Unfortunately, according to Brodu we cannot rely on the testimony of Villars, as he was using and reworking stories of marvels for his own satirical reasons.[8]

Not surprisingly, since the 1950s, Agobard's account has been inflated and twisted in the UFO literature to make it sound more like a visitation from a flying saucer or space ship. Some have reported that these people from the sky had large heads, pointed ears and bulging eyes or were foreign-looking. According to Jacques Bergier's version of the story, the people who were abducted thought that their adventure had taken a short period of time, yet much more time had elapsed on Earth. He thought they had moved at great speed to account for these time differences. Today we would use hypnotic regression to get to the 'truth' of what happened during this missing time.[9]

In the 1960s and 1970s, Agobard was used by Jacques Vallee, John Michell and John Keel to show that abductions have occurred throughout history by beings who are not necessarily from outer space.[10] More recently his story of crops flattened by 'magical storms' has been used to show that crop circles are nothing new and are related to aerial phenomena.[11] Brodu criticized them for ignoring the social and ideological context of Agobard's story, and that there has been too much emphasis on Villars' sensationalistic version of events.

The Demon Ship-Yard

Early Irish monastic literature has several accounts of 'demon ships' in the sky that are regarded as the origin of Agobard's story. The *Book of Leinster* reports that at a royal fair at Teltown (Tailtiu), County Meath the High King Domhnall, son of Murchad in 763 A.D, saw three ships. King Congallach (Congalach) saw one ship in the Tenth Century, according to the *Book of Glendalough*. He saw one of the aerial sailors spear a salmon and 'swim' down to retrieve his catch. The crewman was grabbed but the King ordered him to be released and he swam back to his ship. Bishop Patrick also mentioned the spearing of a salmon in *Mirabilia* ("Wonders"):

> There was once a king of the Scots at a show with a great throng, thousands in fair array. Suddenly they see a ship sail past in the air, And from the ship a man then cast a spear after a fish; The spear struck the ground, and he, swimming, plucked it out.
>
> Who can hear this wonder and not praise the Lord of Thunder?[12]

These Irish manuscripts all refer to a royal show being visited by demon ships, but the chroniclers provide different dates and details.[13] There is a progression from simply seeing ships in the sky to seeing their crewmen spearing fish and 'swimming' down to retrieve the spear. The earlier accounts also refer to thunder in conjunction with these stories and to the grabbing and release of these aerial sailors. Agobard's account changes the spear to an anchor, which set a trend, as there is a whole series of stories that refer to a cloudship that snags its anchor. These again have different dates, locations and minor details according to whichever source you refer to.

A cloud-ship's anchor is said to have been caught on the porch of St. Kinarus's church, Ireland, in 956 A.D. The story is documented in the *Speculum Regale*:

> There happened in the borough of Cloera, one Sunday while people were at mass, a marvel. In this town, there is a church to the memory of St. Kinarus. It befell that a metal anchor was dropped from the sky, with a rope attached to it, and one of the sharp flukes caught in the wooden arch above the church door.

> The people rushed out of the church and saw in the sky a ship with men on board, floating at the end of the anchor cable, and they saw a man leap overboard and pull himself down the cable to the anchor as if to unhook it. He appeared as if he were swimming in water. The folk rushed up and tried to seize him; but the bishop forbade the people to hold the man for fear it might kill him.
>
> The man was freed and hurried up the cable to the ship, where the crew cut the rope and the ship rose and sailed away out of sight. But the anchor is in the church as a testimony to this singular occurrence.

The French chronicler St Peter of Vigeois, claimed that an airship dropped its anchor in the centre of London in 1122. Gervase of Tilbury records a more detailed 'anchor' story in the early 13th century *Otia Imperialia*:

> As people were coming out of church in Britain, on a dark cloudy day, they saw a ship's anchor fastened in a heap of stones, with its cable reaching up from it into the clouds. Presently they saw the cable strained, as if the crew was trying to pull it up, but it still stuck fast. Voices were then heard above the clouds, apparently in clamorous debate, and a sailor came down the cable. As soon as

> he touched the ground the crowd gathered around him, and he died, like a man drowned at sea, suffocated by our damp thick atmosphere. An hour afterwards, his shipmates cut the cable and sailed away; and the anchor they left behind was made into fastenings and ornaments for the church door, in memory of this wondrous event.[14]

The location for this event is often cited as Gravesend or near Bristol. The latter might be because it is confused with another strange story reported by Gervase of Tilbury:

> This port is the one used by the most of those who travel to Ireland. On one occasion a native of that place set sail from that port for Ireland, leaving his wife and family at home. His ship was driven far out of its course to the remote parts of the ocean and there it chanced that his knife fell overboard, as he was cleaning it one day after dinner. At that very moment his wife was seated at table with their children in the house at Bristol, and, behold, the knife fell through the open skylight, and stuck in the table before her. She recognized it immediately, and when her husband came home long afterwards, they compared notes, and found that the time when the knife had fallen from his hands corresponded exactly with that in which it had been so strangely recovered.

The nautical theme of all these stories is not so surprising when we look at the origins and context of these accounts. On a simplistic basis Heaven has always been regarded as amongst or above the clouds. The Christian Bible is full of accounts of extraordinary clouds including the prediction that the Lord would descend from the heavens with the sound of the trumpet of God, "the dead in Christ shall rise first: Then we which are alive and remain shall be caught up together with them in the clouds to meet the Lord in the air: and so shall we ever be with the Lord."[15]

Using references from the *Psalms*, *Genesis* and *Isaiah*, Christians argued that the Earth was flat and that the firmament was a solid roof that stretched over us. Above this roof is a huge water tank from which the angels regulate the fall of rain. It was thought that if you sailed too far you would end up sailing in the ocean above us, as was the implication of the story of the sailor's knife that fell into his home in Bristol. If you could enter heaven above us by such means, then it made sense that you could also stumble into an opening into the underground chambers of hell. The fear of a great opening into hell, located in the Atlantic Ocean far away from Europe, made it difficult for explorers to recruit crews to explore such regions.

In the 20th century, Charles Hoy Fort reintroduced these ideas with his suggestion that there is a Sargasso Sea moving about above us. Such a sea would account for the hundreds of reports of falls of fish, frogs and other animate and inanimate matter that he collected from his extensive search through newspapers and scientific journals.

Enter the Airships

The ancient ships of the sky have a strong connection with the famous airship scare of 1896-1897 that tore through the USA. The vast majority of the sightings were of lights or objects in the sky that the witnesses thought were a fantastic new airship soaring through the air with its proud inventor at its helm. Here again there were reports of unknown aerial craft dropping anchor. At the end of March 1897, they tried to catch someone with their anchor. A farmer called Robert Hibbard, who lived near Sioux City, Iowa, had his clothing hooked by an airship's anchor. After dragging him along the ground for several meters, he escaped when he grabbed a sapling. The anchor flew away with a portion of his trousers as he fell to the ground.[16]

Another anchor turns up in the *Daily Post* (Houston, Texas) of 28 April 1897, which is chillingly like those reported in the Middle Ages:

> Merkel, Texas, April 26 - Some parties returning from church last night noticed a heavy object dragging along with a rope attached. They followed it until in crossing the railroad it caught on a rail. On looking up they saw what they supposed was the airship. It was not near enough to get an idea of the dimensions. A light could be seen protruding from several windows; one bright light in front, like the headlight of a locomotive. After some ten minutes, a man was seen descending the rope; he came near enough to be plainly seen. He wore a light-blue sailor suit, was small in size. He stopped when he discovered parties at the anchor and cut the rope below him and sailed off in a northeast direction. The anchor is now on exhibition at the blacksmith shop of Elliott and Miller and is attracting the attention of hundreds of people.[17]

The Daily Chronicle (Muskegon, Michigan), of 30 April 1897, reported that on the previous evening at 11.30 PM an airship visited the town of Holton:

> It came from the north and descended until it was about 200 feet from the ground, directly over the bridge. It was lighted with electricity and loaded with revellers who were making a good deal of noise.
>
> The music was entrancing; the like of which never was heard in this place. It wasn't long before everybody was on the street to look and listen, many in their nightclothes. Not a few thought the Judgment Day had come. It was 300 feet long, tail about forty feet. Its breadth and depth was about ninety feet. It stayed fifty-five minutes. Its tail commenced whirling and it moved off toward Fremont. But just as it began to move, a grappling hook was let down and caught one of our most truthful citizens who was instantly hoisted on board and carried away. The truthful citizen came back on the 11.30 train from White Cloud and has been talking ever since about aerial navigation.

The glib references to the "truthful citizen" indicate that this story is one of the many tall stories the newspapers used to poke fun at the sightings of the airship. Despite it being true or not it is interesting that this airship was associated with unknown music, Judgment Day, bright electric light and abduction by means of an anchor.

Over the years several authors have considered the stories of ships in the sky, dropping anchor. They are particularly impressed by the Merkel, Texas, case and its close similarity to the accounts in the ancient texts like the *Otia Imperialia*. Abductee Whitley Strieber mentioned these historical cases, adding that the kobold dwarfs who were said to haunt medieval German mines also wore dark blue outfits like the sailor in the Merkel report. Strieber himself had seen small beings with similar outfits and wondered if this was a uniform alien beings used at night time.[18]

The research of Robert Neely deflated these intriguing musings by noting that by the end of April 1897 an article entitled "A Sea Above the Clouds" appeared in large and small newspapers throughout the country. This recounted the drowning sailor from the Thirteenth Century story, and Gervase of Tilbury's airship anchor and knife from the sky stories. As far as Neeley could tell this article first appeared in the *Nebraska State Journal* of 8 March 1897.[19] It is noteworthy that no witnesses in Merkel were actually named or interviewed by the newspapers. Therefore, in all probability an enterprising journalist adapted or invented the Merkel story. The same article could also have been the source for Hibbard's story of being dragged by an anchor, near Sioux City. Daniel Cohen in his book *The Great Airship Mystery* thought it was an obvious tall tale told in a serious manner.[20]

John Keel, in his seminal *UFOs: Operation Trojan Horse,* suggested that we dismiss these stories as 'poppycock' or we have to accept them as true. If we take the latter option, Keel wondered why space ships would need or use primitive anchors?[21] Perhaps they are just poppycock, especially when we look at the famous Hamilton calf-napping case.

Aliens of Whole Cloth

On 19 April, the airship pilots fished for a two-year-old heifer at the Le Roy, Kansas farm of Alexander Hamilton. The craft was seen at 10.30 PM by Hamilton, his son and Gid Heslip. As it slowly descended, they could see it was a 300-foot-long cigar-shaped vessel with a carriage underneath it. The carriage appeared to be made of narrow transparent strips alternating with strips of dark reddish material. A "great turbine wheel" underneath the carriage seemed to be its means of propulsion in the air.

Inside were "six of the strangest beings I ever saw. They were jabbering together but we could not understand a word they said", according to Hamilton's sworn affidavit. Going to a nearby heifer that was tangled in a fence they found "some material" in a slipknot around its neck leading up to the airship. Once the men freed the heifer from the wire fence the airship slowly ascended taking the beast with it. The next day Link Thomas found the hide, legs and head of the heifer on his farm located four miles west of Le Roy.[22]

Cohen showed that the Hamilton story was declared a hoax in 1943 long before ufologists were even aware of it. The editor of *The Farmer's Advocate* admitted that he and Hamilton had concocted the whole thing.[23]

The diligent research of Thomas Bullard revealed that the *Atchison County Mail* published as early as 7 May, 1897, a letter from Hamilton confessing that his story was a hoax explaining that "whenever I get a chance to help Kansas in any way I always do it..."[24] In the 1960s the original news reports were re-discovered by ufologists and repeated in countless UFO books. It was not until 1976 that anyone explored the possibility that it was a hoax. It was found that in his local community Hamilton was known for his whopping stories. He and his friends had a weekly liars club where they tried to top each other's tall stories, and the calf-napping tale was their most famous achievement. Jerome Clark published these findings in *Fate* magazine, but the Hamilton calfnapping is still repeated in many UFO books as a fact.[25]

The Hamilton calf-napping incident sounds reasonable compared to other stories that circulated in the newspapers of the period, which were also backed by the testimony of 'reliable and respectable' witnesses. Some people claimed to meet the pilots of the airships as they re-stocked their craft or made repairs. Some, like the contactees of the 1950s, got invited onboard the aerial vessel where they were shown its technological marvels.

Mars was also said to have been the home planet of an alien whose craft crashed into a windmill at Aurora, Texas on 17 April 1897. The pilot and his craft were badly mangled, leaving tons of aluminium and silver wreckage. Papers containing indecipherable hieroglyphics were found on the body of the alien. In the 1960s, when the case was reinvestigated, it was discovered that there had never been a windmill at this location. In the 1970s several eyewitnesses to the crash were located at a Nursing Home, where they claimed that they never saw the airship crash and that the newspapers had made-up their stories. This was almost certainly a hoax, yet this has not stopped UFO researchers from searching for the Martian's grave and fragments of his spaceship. This indicates that ufological 'holy grails,' such as Roswell, are by no means a new phenomenon.[26]

Flexible Phantoms

Much of the early history of ufology might have been lost to us entirely if it was not for the research of Charles Hoy Fort (1874-1932). His years of diligent noting of these mysteries created a body of work that was not very popular at the time, but it has had a long-lasting and insidious impact on science fiction, ufology and paranormal literature. He often speculated about the possibility that we were the property of extraterrestrials. In his, book *Lo!* he mused that:

> It may be that if beings from somewhere else would seize inhabitants of this earth, wantonly, or as a matter of scientific research, the preference would be...remote from observations by other humans of this earth.

The many accounts of ships, armies and visionary figures seen in the sky that are often listed by UFO writers as evidence of UFO events in history can often be explained as sightings of mundane mirages, meteors, ball lightning, misperceptions and the like. As in the instance of Agobard's story, they gradually change and alter out of all recognition from their original sources.

In the last few centuries, sightings of things in our skies have changed in accordance with our own invention and development of balloons, airships, aircraft and space vehicles. The UFOs seem to reflect or even anticipate our technological creations. This has led ufologists to postulate the concept of cultural tracking.

The theory that the aliens shape themselves and their craft in the manner that suits the cultural and social conditions of the time has several flaws. The main assumption is that the flying saucers and their grey occupants are the 'real' manifestation of these objects and that all the previous manifestations were manipulated by the aliens. This has led to the concept of the aliens as being cosmic tricksters, who will always remain tantalizingly beyond our human grasp.

Alternatively, the cultural tracking argument can be turned upside down: we interpret things in the sky according to our own social, cultural and psychological expectations. This would explain why we collectively and individually see things in the sky such as dragons, flying saucers, angels or foo fighters according to our preconceptions rather than any external alien forces. A case in point is the vision of Ezekiel that has been interpreted in our own time as a spaceship, and in the past as a flying machine, airship, locomotive and chariot.[27] The Biblical description has not changed but our own pre-conceptions of what it means have. Our minds are the flexible phantoms, not the aliens.

Nigel Watson is the author of *The Alien Deception* (2009), *Portraits of Alien Encounters* (1990), *Supernatural Spielberg* (with Darren Slade, 1992) and editor/writer of *The Scareship Mystery* (2000). He has written for numerous publications, including *Paranormal Magazine, Fortean Times, Beyond, Wired, History Today, Flipside, Strange Magazine, Magonia, The Unexplained, Flying Saucer Review, Alien Worlds,* and *UFO Magazine.* He produces the media website Talking Pictures (www.talkingpix.co.uk). He has a BA degree in Psychology (Open University) and a BA (Hons) degree in Film and Literature (University of Warwick). He lives in Plymouth, UK, with his wife and three children.

Endnotes

David Luke - Disembodied Eyes (p. 9)

Notes:

1. Rodriguez, Marko A., "A Methodology for Studying Various Interpretations of the N,N-dimethyltryptamine-Induced Alternate Reality", Journal of Scientific Exploration, 2006. Grossly simplified, Rodriguez proposes obtaining from the entities solutions to complex mathematics puzzles that the DMT experient does not know. Regrettably, this ingenious method for testing the reality of DMT entity encounters is subject to a number of flaws, such as the huge assumptions involved in expecting our supposed entities to be hyper-intelligent and/or have the desire to cooperate. The most crippling problem for his test, however, is what is known as the superpsi hypothesis – an issue long-proven difficult to surmount in parapsychological attempts to validate the existence of discarnate entities considered to be spirits of the dead (e.g. those apparently communicating via trance mediums). The problem is that, because clairvoyance, telepathy, and precognition (collectively called psi) have no theoretical (or even apparent) limits, it always remains a possibility that any information provided by ostensibly discarnate entities may actually be due to the super-psi of the person (e.g., the medium) receiving the information (see Braude 2002 for a comprehensive discussion). However, such an issue doesn't carry quite as much currency with my perspective on investigating entities in this manuscript.

2. Acronym (**fo**lklore **a**nd **my**thology, **cu**ltural **st**udies **a**nd **r**elated **d**isciplines) courtesy of foamy custard chef, Bob Trubshaw. For information, see http://www.indigogroup.co.uk/foamycustard.

3. "You get elves, everybody does." Audio remix at Trip Receptacles, http://www.nvo.com/cd/trip.

4. Nebesky-Wojkowitz, R. 1956. *Oracles and Demons of Tibet: The Cult and Iconography of the Tibetan Protective Deities.* Oxford University Press.

5. Kent, J. 2005. "The Case Against DMT Elves," in C. Pickover, *Sex, Drugs, Einstein, and Elves.* Smart Publications. pp. 102–105.

6. Turner, D.M. 1995. "Exploring Hyperspace," *The Entheogen Review* 4(4): 4–6.

7. Meyer, P. 1993. "Apparent Communication with Discarnate Entities induced by Dimethyltryptamine (DMT)," in T. Lyttle (ed.), *Psychedelic Monographs & Essays* 6: 28–67. PM&E Publishing Group.

8. Luke, D.P. and M. Kittenis 2005. "A Preliminary Survey of Paranormal Experiences with Psychoactive Drugs," *Journal of Parapsychology* 69(2): 305–327.

9. http://www.erowid.org/experiences/exp.php?ID=6217

10. By "radical empirical" I refer to the term as championed by William James (2003/1912), which posits that standard scientific empiricism tries to reduce experience to bare sensations at the expense of prior reasoning, intuition, revelation, or meaning. James argued that we see the world in terms of meaning and the actual connections made between phenomena, so introspection of experience is as legitimate a scientific enterprise for studying one's experience and the contents of one's own mind as is the mere observation of the sensation of experience (i.e., empiricism).

11. https://dmt-nexus.com/forum/default.aspx?g=posts&t=141

12. Sheldrake, R. 1988. *The Presence of the Past.* Fontana/Collins.

13. Strassman, R. 2001. *DMT: The Spirit Molecule: A Doctor's Revolutionary Research into the Biology of Near-Death and Mystical Experiences.* Park Street Press.

14. Incidentally, when Alex Grey was asked about the prevalence of disembodied eyes in psychedelic visions he merely said that they represent infinite awareness. However, he recalled that a woman who had seen his *Dying* painting had reported once traveling down a tunnel of eyes during a near-death experience (Hanna 1998).

15. Jackson, N. and M. Howard 2000. *The Pillars of Tubal Cain.* Capall Bann Publishing.

16. Boyce, M. 1975. *History of Zoroastrianism* Vol. I. Brill.

17. Crowley, M. 1996. "When the Gods Drank Urine: A Tibetan Myth May Help Solve the Riddle of Soma, Sacred Drug of Ancient India," *Fortean Studies* 3.

18. Crowley, M. 2005 (Dec. 5). Personal communication.

19. Rudgley, R. 2000. *The Encyclopedia of Psychoactive Substances.* Thomas Dunne.

20. Frank Olsen was a U.S. Army officer who was unwittingly dosed with LSD by the CIA in 1953. A few days later – under CIA supervision – he plummeted to his death from a hotel window. It was alleged that he jumped, and the drugging incident was covered up for over twenty years as part of the secret MKULTRA operation. Forensic evidence from 1994 added a new twist, strongly suggesting that Olsen was pushed: a murder, rather than a suicide.

21. McKenna, T. 1992. *Food of the Gods: The Search for the Original Tree of Knowledge.* Bantam.

Richard Andrews - Closing the Circle (p. 41)

Notes:

1. Original image courtesy of Steve Alexander (www.temporarytemples.co.uk). Euler's identity image from Wikimedia Commons (http://commons.wikimedia.org/wiki/File:Euler%27s_formula.svg), licensed under Creative Commons Attribution-Share Alike 3.0 ("Originally created by gunther using xfig, recreated in Inkscape by Wereon). Additional graphic design by Richard J. Andrews.
2. Professor Keith Devlin, Stanford University, as quoted in Paul J. Nahin (2006). *Dr. Euler's Fabulous Formula: Cures Many Mathematical Ills.* Princeton University Press.
3. http://logosmythos.net/wilton_windmill_crop_circle_may_2010.html
4. Original Wilton Windmill crop circle illustration by Andreas Müller (http://www.kornkreise-forschung.de/). Additional graphic design by Richard J. Andrews.
5. Rob Irving, "Art and Artifice", from http://www.circlemakers.org/art_and_artifice.html
6. Rob Irving, John Lundberg and Mark Pilkington. (2006). *The Field Guide: The Art, History and Philosophy of Crop Circle Making.* Strange Attractor Press.
7. George P. Hansen. (2001). *The Trickster and the Paranormal.* Xlibris Corporation.
8. Allan Brown and John Michell. (2005). *Crooked Soley: A Crop Circle Revelation.* Roundhill Press.
9. Jim Schnabel. (2002). *Round in Circles: Poltergeists, Pranksters, and the Secret History of the Cropwatchers.* Prometheus Books.
10. Jeffrey J. Kripal. (2010). *Authors of the Impossible: The Paranormal and the Sacred.* University of Chicago Press.
11. Dez Vylenz. (2008). *The Mindscape of Alan Moore.* (DVD). Shadowsnake Films.

Martin Shough - Return of the Flying Saucers (p. 69)

Notes:

1. Arnold's 1952 book recalls that he used this last simile "at the time" in 1947: "They flew in a definite formation, but erratically. As I described them at the time, their flight was like speed boats on rough water…" (*The Coming of the Saucers* p.11) The earliest published source I have so far been able to identify for the speed-boat simile is the April 1950 interview with broadcaster Ed Murrow, but it may occur in an early press story unknown to me.

2. Radio broadcast of reconstructed telephone interview with Kenneth Arnold by Ed Murrow, April 07 1950.

3. 'Links Flying Disks with Mystery Sub', UP, Boise, Ida. April 10 1950

4. Long, Greg., 'Kenneth Arnold: UFO-Pioneer', *MUFON Journal*, Nov. 1981 p.7

5. 'Harrassed Saucer-Sighter Would Like to Escape Fuss,' Boise, Idaho *Statesman*, June 27, 1947

6. 'Cub Scribe Broke Story', by James Long, Staff Writer, *Oregon Journal* Oct 17 1975

7. It also contains the phrase repeated in many places by Arnold, "flat like a pie pan" (and, uniquely, the odd phrase "somewhat bat-shaped" - see later).

8. Lagrange, P., 'A Moment in History: An Interview with Bill Bequette', IUR Vol. 23, No. 4, Winter 1998. According to historian Loren Gross (*UFOs: A History*, privately produced, Fremont, Ca., 1988, Vol. 1. p.7), Arnold told Nolan Skiff that the "the 'missiles' travelled like a flat rock bounced across the surface of water, a rising and falling motion." This would be the natural form of such a simile intended only to illustrate motion. It is a commonplace. Everyone has skipped stones, whereas "skipping *saucers*" is on the face of it an unlikely activity, so Gross's account appears plausible. It is called in question because Gross adds that Bequette used this simile to invent the term "flying saucers" for his AP wire story. He did not. On the other hand, a similar expression was attributed to Arnold by Lagrange after having interviewed Bequette: "they look like pebbles [flat stones] or plates: flat, rounded at the front, triangular at the rear" ("On dirait des galets ou des assiettes: plats, arrondis a l'avant. l'arrere triangulaire") (See: Lagrange, P., 'How it All Began', *Anomalies* #3 1997 p.27)

9. Story, R., "Mammoth Encyclopedia of Extraterrestrial Encounters", 2001, p100. But note that Story incorrectly states that this identifies Bequette as "the man who coined the term 'flying saucer'". We can be certain that this is not true, both because of Bequette's explicit disclaimers and because we can prove that the term does not appear in the articles or wire stories written by Bequette. Evidently Bequette was giving to Story the same concession he had offered previously to Lagrange - that, to give Arnold the benefit of the doubt, the phrase "saucer-like" might, rather than being an exact quote, have been Bequette's own, but one "based on Arnold's description [of shape]".

10. Norman (Oklahoma) *Transcript* June 26; see also Cleveland (Ohio) *Press*, Friday, June 27, 1947 and other papers

11. Chicago *Daily Tribune*, June 26 1947 p.1

12. Live broadcast interview with Kenneth Arnold by Ted Smith, KWRC Pendleton, June 26 1947, rediscovered by Pierre Lagrange, see: http://www.ufologie.net/htm/arnoldrepbequette.htm

13. Arnold, K., report to the Commanding General, Army Air Forces, Wright Field, July 08 1947.

14. Email to the author from Barry J. Greenwood, Dec 28 2009

15. http://www.thefreedictionary.com/BAT

16. Brown and Davidson met with Arnold and Eastern Airlines pilot Capt E.J.Smith in Tacoma to be informed of the two men's discoveries about the claimed saucer sighting at Maury Island (see later). They showed Arnold a drawing or drawings based on photos taken by William.Rhodes in Phoenix, which the AAF believed "genuine". Arnold then excitedly recalled that one of his nine objects had looked just the same. The first dated representation of a crescent is a drawing done by Arnold a couple of weeks later, shortly before Aug 19 1947, on a copy of his AAF letter given to the FBI. Arnold told the agent on that date that Lt Brown had described one of the Rhodes photos to him as resembling a "flying wing".

17. Arnold, K., "How It All Began" in Fuller, Curtis G., *Proceedings of the First International UFO Congress [1977]*, Warner, 1980, pp. 17-29.

18. He said that they did appear "completely round" at moments when the sun reflected specularly off their top sides, but the true, slightly eccentric, shape was revealed in silhouette against the snow (letter to the Commanding General, Wright Field, July 08 1947).

19. Portland *Oregonian*, Sunday, June 29, 1947, p. 24.

20. 'Are They Jet Propelled? Oregon Man Sights Disks at Close Range', *Oregon Daily Journal,* June 28 1947

21. Portland *Oregonian*, Sunday, June 29 1947, p. 26

22. Arnold, K. & Ray Palmer, *The Coming of the Saucers,* Amherst Press, Wisconsin 1952, p.17

23. 'Harassed Saucer-Sighter Would Like to Escape Fuss,' Boise, Idaho *Statesman*, June 27, 1947

24. Arnold, K., "How It All Began" in Fuller, Curtis G., *Proceedings of the First International UFO Congress [1977]*, Warner, 1980, pp. 17-29

25. See http://www.martinshough.com/aerialphenomena/Arnold%20analysis2.pdf Appendix *4.e* for a timeline of the imagery used by Arnold and others between 1947 and 2009

26. http://www.martinshough.com/aerialphenomena/Arnold%20analysis2.pdf

27. Various emails from Joel Carpenter, V-J. Ballester Olmos, Mike Tarbell and others.

28. Arnold, K., & Ray Palmer, *The Coming of the Saucers*, Amherst Press, Wisconsin, 1952.

29. Keel, J. A., 'The Maury island Caper,' in: Evans & Spencer (eds) *UFOs: 1947-1987, The 40-year Search for an Explanation*, Fortean Tomes, London 1987 pp.40-43

30. Garrison, G., *On the Trail of the Assassins,* Sheridan Square Press, 1988, pp.207-213 & photos facing p.176.

31. Arnold's first-person account appears in the 1952 book co-written with Ray Palmer. According to CUFOS archivist & historian Frank Reid , Arnold recorded his contributions on a dicta-belt machine and sent the belts to Palmer who transcribed them for the manuscript (emails from Mark Rhodegier & Frank Reid, 17.07.2010 *et seq.*). Palmer's editorial touch seems to have been a light one. The narrative sounds like Arnold's "voice" to me.

32. Email from from V-J. Ballester Olmos, 18.07.2010

33. A claim (*ibid.*) that Arnold had a history of "weird experiences" going back to childhood appears to be limited to his 1981 remark to Greg Long about a globe of light said to have been seen at the death of a relative by members of his family, including himself at the age 7. A sixty-year-old family story of this character cannot really be given much weight as evidence of an abnormal predisposition to the bizarre on the part of the 32-year-old pilot. See: Long, Greg., 'Kenneth Arnold: UFO-Pioneer', *MUFON Journal,* Nov. 1981 p.7

34. Email from V-J.Ballester-Olmos, 18.07.10. See: V-J., & Barry Greenwood, *The Year 1947 in Photos* (in press)

35. Notarised affidavit of David M. Johnson, Boise, Idaho, July 12 1947, exhibited to Lt. Frank M. Brown, S/A CIC, 5th Air Force. (ATIC files)

36. Arnold said the objects "veered sharply away from me and to *their* right" which is clearly a slip.

37. e.g. Kottmeyer, M., 'Resolving Arnold - Part 2: Guess Again', *REALL Newsletter,* Volume 5, Issue 7, July 1977. http://www.reall.org/newsletter/v05/n07/resolving-arnold-part-2.html

38. Easton, J., Voyager Newsletter #10, Apr 07 2000 see: http://www.ufoupdateslist.com/2000/apr/m07-018.shtml

39. Arnold, K. & R. Palmer, *The Coming of the Saucers,* Amherst Press, Wisconsin, 1952 pp.25-7

40. McDonald. Dr. J.E., notes of telephone interview with Arnold, Nov 19 1966 (CUFOS files)

41. Notes of telephone interview with Arnold by Dr Richard Read, University of Seattle, March 11 1965 (CUFOS files)

42. *Ibid.*

43. Informal colloquium held at the 1977 International UFO Congress; email to the author from CUFOS archivist Frank J. Reid, 28.07.2010.

44. McDonald. Dr. J.E., notes of telephone interview with Arnold, Nov 19 1966 (CUFOS files)

Nick Redfern - In Search of the Last Mammoth (p. 101)

Sources:

Interview with Jill O'Brien, January 9, 2009;
"The Mammoth Still in the Land of the Living", Edward Newman, the *Zoologist*, October, 1873;
Argus, February 1888;
Daily Gazette, August 23, 1873;
On the Track of Unknown Animals, Bernard Heuvelmans, Hill & Wang, 1959;
"Tales of Living Mammoths", Lindsay Selby, *Still on the Track*,
http://forteanzoology.blogspot.com/2010/05/lindsay-selby-tales-of-living-mammoths.html
"A Mammoth Hoax", Chad Arment, at the Strange Ark website:
http://www.strangeark.com/bfr/articles/mammoth-hoax.html

Blair MacKenzie Blake - Dark Seed (p. 117)

Notes

1. Called *Po Tolo* by the Dogon, Sirius B is identified with the tiniest seed known to them, the botanical Latin name for which is *Digitaria exilis*.
2. Knowing that these shamanic flights are often induced by botanical aids, in ancient Egypt this may have been the indigenous Blue Water lily [*Nymphaea caerulea*] depicted in magical papyri and on funerary stelae, but could have been some other extract of wood, root, or seed having pronounced psychoactive effects.
3. The name Oge-Make is itself intriguing, as it could be related to the Dogon's "Ogo" – a word that figures prominently in their creation myth, representing humankind... "Ourselves", as Temple reminds us, "in all our cosmic impurity."
4. That is if we discount that other nontelescopic observer who possibly knew of the existence of Sirius B. Here I'm referring to the French philosopher and satirist, Voltaire, whose remarkable story *Micromegas* (1752) concerns a giant from Sirius and his dwarf companion who travel to earth to impart wisdom to a certain priest.

Robert M. Schoch - Moving the Moai (p. 135)

Bibliography:

Anonymous/BBC News. "'Hobbit' joins human family tree," Article last updated 27 October 2004 Available at http://news.bbc.co.uk/2/hi/science/nature/3948165.stm Accessed 29 July 2010.

Bahn, Paul, and John Flenley. *Easter Island, Earth Island.* London: Thames and Hudson, 1992.

Barthel, Thomas S. *The Eighth Land: The Polynesian Discovery and Settlement of Easter Island.* Honolulu: The University Press of Hawaii, 1978.

Bettocchi, Lorena. "Rongorongo: Les Écritures de L'Ísle de Pâques," *Kadath: Chroniques des Civilisations Disparues,* no. 105, pp. 4-79 (2009).

Braude, Stephen E. "The Fear of Psi: It's the Thought that Counts," *Darklore* vol. 2, pp. 99-111 (2008).

Brown, J. Macmillian. *The Riddle of the Pacific.* London: T. Fisher Unwin, 1924.

Brown, P., T. Sutikna, M. J. Morwood, R. P. Soejono, Jatmiko, E. Wayhu Saptomo, and Rokus Awe Due, "A new small-bodied hominin from the Late Pleistocene of Flores, Indonesia," *Nature* vol. 431, pp. 1055-1061 (28 October 2004). See http://www.nature.com/nature/journal/v431/n7012/full/nature02999.html Accessed 29 July 2010.

Chauvet, Stephen. *La Isla de Pascua y sus Misterios.* Santiago: Zig-Zag, 1946.

de Laat, M. *Words Out of Wood: Proposals for the decipherment of the Easter Island script.* Delft, The Netherlands: Eburon, 2009.

Fischer, Steven Roger. *Rongorongo: The Easter Island Script.* Oxford: Clarendon Press, 1997.

Fischer, Steven Roger. *Island at the End of the World: The Turbulent History of Easter Island.* London: Reaktion Books, 2005.

Flenley, John, and Paul Bahn. *The Enigmas of Easter Island: Island on the Edge.* Oxford and New York: Oxford University Press, 2002/2003. [New version of Bahn and Flenley, 1992.]

Fox, C. E. *The Threshold of the Pacific.* London: Kegan Paul, Trench, Trubner and Co., 1924. [Also published in the U.S.: New York: Alfred A. Knopf, 1925.]

Haun, Beverley. *Inventing 'Easter Island.'* Toronto: University of Toronto Press, 2008.

Heyerdahl, Thor. *Aku-Aku: The Secret of Easter Island.* Chicago: Rand-McNally, 1958.

Heyerdahl, Thor, and Edwin N. Ferdon, Jr., editors. *Reports of the Norwegian Archaeological Expedition to Easter Island and the East Pacific, Volume 1: Archaeology of Easter Island.* Chicago: Rand McNally, 1961.

Heyerdahl, Thor, and Edwin N. Ferdon, Jr., editors. *Reports of the Norwegian Archaeological Expedition to Easter Island and the East Pacific, Volume 2: Miscellaneous Papers*. Chicago: Rand McNally, 1965.

Heyerdahl, Thor. *Easter Island: The Mystery Solved*. New York: Random House, 1989.

Jones, Alexander, general editor. *The Jerusalem Bible, with Abridged Introductions and Notes*. London: Darton, Longman and Todd, 1968. [This is the edition of the scriptures from which the quotation from St. John was taken.]

Kulagin, Ing. V. V. [husband of N. S. Kulagina]. "Nina S. Kulagina," in *Symposium of Psychotronics, Prague, Czechoslovakia*, 25 September 1970 (Dr. Zdenek Rejdak, chief editor), published in *Journal of Paraphysics*, vol. 5, nos. 1 & 2 (double issue), pp. 1- 62 (Kulagin article, pp. 54-62), Paraphysical Laboratory, Downton, Wiltshire, England, 1971.

Lee, Georgia. *The Rock Art of Easter Island: Symbols of Power, Prayers to the Gods*. University of California, Los Angeles. The Institute of Archaeology. *Monumenta Archaeologica* 17, 1992.

London Dialectical Society. *Report on Spiritualism, of the Committee of the London Dialectical Society, together with the Evidence, Oral and Written, and a Selection from the Correspondence*. London: J. Burns, 1873.

Mazière, Francis. *Mysteries of Easter Island*. New York: W. W. Norton, 1968. [French edition: *Fantastique Île de Pâques*. Paris: Robert Laffont, 1965.]

Magli, Giulio. *Mysteries and Discoveries of Archaeoastronomy: From Giza to Easter Island*. New York: Copernicus Books, 2009.

McLaughlin, Shawn. *The Complete Guide to Easter Island, second edition*. Los Osos, California: Easter Island Foundation, 2007.

Métraux, Alfred. "Easter Island," *Annual Report of the Board of Regents of the Smithsonian Institution showing the operations, expenditures, and condition of the institution for the year ended June 30, 1944*. Washington, D. C.: Government Printing Office, 1945, pp. 435-451 plus 4 plates.

Métraux, Alfred. "Mysteries of Easter Island," *Yale Review*, vol. 28, pp. 758-779 (1939).

Métraux, Alfred. *Easter Island: A Stone-Age Civilization in the Pacific*. New York: Oxford University Press, 1957.

Peiser, Benny. "From Geonocide to Ecocide: The Rape of Rapa Nui," *Energy and Environment*, vol. 16, nos. 3 & 4, pp. 513-539 (2005). Available at http://www.staff.livjm.ac.uk/spsbpeis/EE%2016-34_Peiser.pdf Accessed 21 Feb. 2010.

Routledge, Mrs. Scoresby [Katherine Pease Routledge]. *The Mystery of Easter Island: The Story of an Expedition*. London: Hazell, Watson and Viney, 1919.

Schoch, Robert M., and Logan Yonavjak, editors and compilers. *The Parapsychology Revolution: A Concise Anthology of Paranormal and Psychical Research*. New York: Jeremy P. Tarcher/Penguin, 2008.

Sereno, Martin I. "Language Origins Without the Semantic Urge," *Cognitive Science Online* 3.1: 1–12. Available at http://cogsci-online.ucsd.edu/3/3-1.pdf (2005) Accessed 29 July 2010.

Soza, Felipe L. *Easter Island, Rapa Nui.* Santiago: Editorial S&E S.A., 2007.

Thomson, William J. "Te Pito Te Henua, or Easter Island," *Report of the U. S. National Museum, Under the Direction of the Smithsonian Institution, for the Year Ending June 30, 1889,* 1891, pp. 447-552.

Thorpe, Nick. *Eight Men and a Duck: An Improbable Voyage by Reed Boat to Easter Island.* New York: The Free Press, 2002.

Van Tilburg, Jo Anne. *Easter Island: Archaeology, Ecology, and Culture.* Washington, D. C.: Smithsonian Institution Press, 1994.

Verlinde, Erik. "On the Origin of Gravity and the Laws of Newton," Paper dated 6 January 2010, 29 pages. Available at http://arxiv.org/PS_cache/arxiv/pdf/1001/1001.0785v1.pdf Accessed 28 July 2010.

Wolff, Werner. *Island of Death: A New Key to Easter Island's Culture through an Ethno-psychological Study.* New York: J. J. Augustin, 1948.

Notes:

1. Peiser, 2005.
2. Mazière, 1968.
3. Mazière, 1968, p. 133.
4. Routledge, 1919, p. 198, italics in the original.
5. Wolff, 1948, p. 157, citing Métraux, 1939, p. 771.
6. Schoch and Yonavjak, 2008, and references cited therein.
7. Verlinde, 2010, p. 1.
8. London Dialectical Society, 1873; Braude, 2008.
9. Schoch and Yonavjak, 2008, and references cited therein.
10. Kulagin, 1971, pp. 57-58.
11. Wolff, 1948, p. 156, citing Fox, 1924/1925, p. 338.
12. Brown, 1924, p. 279, italics in the original.
13. Quoted in Thomson, 1891, p. 462.
14. Anonymous/BBC, 2004.
15. Brown, et al., 2004.
16. Sereno, 2005, p. 9.

Neil Arnold - The Mill Race Monster (p. 157)

Sources:

With many thanks to Chris McDaniel for bringing this fascinating legend to my attention.

http://en.wikipedia.org/wiki/Mill_Race_Park
Milwaukee Journal – May 26th 1983: "Chicago Angler Lands Millrace Monster"
The Argus Press – February 21st 1984: "Monster Fish Makes Life Rough for Migrating Fowl"
The Republic – November 2nd 1974: "Monster – Women Report Seeing Beast in Mill Race"
The Republic – November 6th 1974: "Mill Race Makes Appearance After Voting ?"
The Republic – November 8th 1974: "Mill Race Monster Hunt Out of Hand'
The Republic – November 9th 1974: "Green Monster No Danger, People Are"
The Republic – November 11th 1974: "No Week End Monster Sightings"
The Republic – November 12th 1974: "Mill Race Park Closed Nightly"
The Republic – October 31st 2003: "Residents Laugh About '74 'Monster Hoax"

Erik Davis - Calling Cthulhu (p. 167)

Notes:

1. "The Haunter of the Dark," *The Best of H.P. Lovecraft: Bloodcurdling Tales of Horror and the Macabre* (New York: Ballantine, 1963), p. 220.

2. Gilles Deleuze and Felix Guattari, *A Thousand Plateaus* (Minnesota: University of Minnesota Press). "The anomalous is neither an individual nor a species; it has only affects, it has neither familiar or subjectified feelings, nor specific or significant characteristics. Human tenderness is as foreign to it as human classifications. Lovecraft applies the term 'Outsider' to this thing or entity, the Thing which arrives and passes at the edge, which is linear yet multiple, 'teeming, seething, swelling, foaming, spreading like an infectious disease, his nameless horror.'" (244-5) "What we are saying is that every animal is fundamentally a band, a pack...It is at this point that the human being encounters the animal. We do not become animal without a fascination for the pack, for multiplicity. A fascination for the outside? Or is the multiplicity that fascinates us already

related to a multiplicity dwelling within us? In one of his masterpieces, H.P. Lovecraft recounts the story of Randolph Carter, who feels his 'self' reel and who experiences a fear worse than that of annihilation: 'Carters of form both human and nonhuman, vertebrate and invertebrate… etc."(240)

3. Ibid., p. 202

4. One notable exception to this is *The Starry Wisdom,* D.M. Mitchell's recent British collection of Mythos tales and art that bristles with experimental and chthonic magick (London: Creation Books, 1994). Contributors range from J.G. Ballard and William S. Burroughs to Grant Morrison and Alan Moore, two of today's most intelligent comic book writers. Though the bios make no mention of the fact, a number of the contributors, including Don Webb, are solid occultists. On the other hand, the author of one of the more perverse tales is Robert M. Price, a New Testament scholar and pastor of the First Baptist Church of Montclair, New Jersey.

5. Fritz Leiber, Jr., "A Literary Copernicus," in S.T. Joshi, et., *H.P. Lovecraft: Four Decades of Criticism* (Athens, Ohio: Ohio University Press, 1980), p. 51.

6. S.T. Joshi, *H.P. Lovecraft* (Mercer Island, Wash.: Starmbat House, 1982), p. 58.

7. He was also the first modern occultist to anticipate Lovecraft by trafficking with explicitly extraterrestrial entities.

8. See Nadia Choucha's *Surrealism and the Occult* (Vermont: Destiny Books, 1991), pp. 51-74.

9. Colin Wilson, "Introduction," in George Hay, ed., *The Necronomicon: The book of Dead Names* (London: Corgi, 1978), p. 35.

10. The Necronomicon FAQ Version 2.0, written and compiled by Kendrick Kerwin Chua (kchua@unf6.cis.unf.edu), May 1994.

11. Anton LaVey, *The Satanic Rituals* (New York: Avon, 1972).

12. Private email correspondence.

13. "The Whisperer in Darkness," op. cit *Best of…*, p. 162.

14. Joshi, p.32.

15. The imaginative production of a role-playing game's "virtual reality" is not dissimilar to ceremonial magic. RPGs consist of a defined space and time; random dicethrows; source books; a game system that organizes numbers, rules and characteristics; and, most importantly, the active imagination of the players. Organized magical rituals follow a superficially similar scheme: a coherent and graded system carves out a performative space and time wherein divinatory randomness, data from occult cookbooks, and symbolic networks of numbers and signs come to life within the active imagination of the mage, a "playful" imagination that calls forth – and even becomes – the autonomous "characters" of the gods.

From ancient ritualistic board games to the Golden Dawn's Enochian Chess, games have been vehicles for magic and divination. Recently Chaosium, the publisher of *Call of Cthulhu,* has cemented the connection with *Nephilim*, the first explicitly occult role-playing game. You play one the Nephilim, a race of magical beings who battle secret societies and must take over living human hosts in their quest for enlightenment. Already banned in some stores, *Nephilim* artfully and systematically melds deep and genuine esoteric arcana (kabbalah, alchemy, astrology) with an original variant of the occult conspiracy theory of history.

16. Following the logic of the Freudian slip, Spare workings such as sigil magic must be "forgotten" in order for their encoded intention to "break through" into reality.

17. August Derleth collected Lovecraft's letters about his dreams in *Dreams and Fantasies* (Suak City, Wis.: Arkham House, 1962).

18. op. cit, Chua.

19. Quoted in Roland Edighoffer,"Rosicrucianism," in Antoine Faivre and Jacob Needleman, eds., *Modern Esoteric Spirituality* (New York: Crossroads, 1992), p. 198.

20. Quoted in S.T. Joshi, "Afterword," in *H.P. Lovecraft, The History of the Necronomicon* (West Warmick, R.I.: Necronomicon Press, 1980), p.9.

Greg Taylor - How Martin Gardner Bamboozled the Skeptics (p. 195)

Notes:

1. Marcello Truzzi, "On Pseudo-Skepticism" http://www.anomalist.com/commentaries/pseudo.html

2. Cited in George Hansen, *The Trickster and the Paranormal*

3. http://www.nytimes.com/2006/08/20/books/review/20Gottlieb.html. Joe Nickell's review of *Ghost Hunters* in *The Skeptic* (November/December 2006) also takes Blum to task for not having read Martin Gardner's analysis of Mrs Piper.

4. Martin Gardner, *The Night Is Large: Collected Essays, 1938-1995*

5. *Proceedings of the SPR* Volume IV, 1887

6. The errors can be found in *Are Universes Thicker than Blackberries* – I have been unable to check if the same mistakes appear in the earlier publication of this essay, in *An Encyclopedia of the Paranormal.*

7. F.W.H Myers, Introduction to "A Record of Observations of Certain Phenomena of Trance", in *Proceedings of the SPR* Volume VI, p. 438.

8. Ibid. p. 441

9. Ibid.

10. Ibid.

11. Richard Hodgson, "A Record of Observations of Certain Phenomena of Trance", *Proceedings of the SPR* Volume VIII.

12. James Hyslop, "President G. Stanley Hall's and Dr. Amy E. Tanner's Studies in Spiritism", in the *Journal of the ASPR* Volume V.

13. William Romaine Newbold, "A Further Record of Observations of Certain Phenomena of Trance, Part II", *Proceedings of the SPR* Volume XIV.

14. Frank Podmore, "Discussion of the Trance-Phenomena of Mrs. Piper", *Proceedings of the SPR* Volume XIV.

15. Ibid.

16. Banachek, *Psychophysiological Thought Reading*

17. Sir Oliver Lodge, "A Record of Observations of Certain Phenomena of Trance", in *Proceedings of the SPR* Volume VI, p. 451.

18. Amy Tanner, *Studies in Spiritism*, p. 45.

19. Frank Podmore, "Discussion of the Trance-Phenomena of Mrs. Piper", *Proceedings of the SPR* Volume XIV.

20. *Journal of the SPR*, Volume VIII.

21. F.W.H Myers, Introduction to "A Record of Observations of Certain Phenomena of Trance", in *Proceedings of the SPR* Volume VI, p. 439.

22. Though this procedure of 'talking to the hand' appears bizarre, to the point of being laughable, it seems likely that it arose as a very practical solution to the problem of having simultaneous communication with two 'spirits' – it was to specify which 'spirit' was being addressed. Sitters spoke to the hand to address the one communicating with writing, or spoke to Mrs. Piper's face to address the 'spirit' who was talking.

23. *Proceedings of the SPR* Volume XXV, pp. 102-108

24. James Hyslop, "President G. Stanley Hall's and Dr. Amy E. Tanner's Studies in Spiritism", in *Journal of the ASPR* Volume V.

25. Ibid.

26. Ibid.

27. Stanley Hall, Introduction to Amy Tanner, *Studies in Spiritism*, p. xxxii

28. *Proceedings of the SPR* Volume XXV, pp. 102-108

29. Amy Tanner, *Studies in Spiritism*, p. 202

30. Ibid, p. 270

31. Ibid. p. 246

32. An immediate sign that we should take care in accepting Gardner's opinion is his use of loaded language, which betrays his bias. For instance, he says that "Mrs. Piper was tall, stout, and handsome, with blue eyes and brown hair, good-

natured, self-possessed, matronly, modest, and *shrewd*". This kind of comment makes one wonder whether, when Gardner says "a clock was *kept* illuminated after the room was darkened," he meant a clock *stayed* illuminated. He comments that this was so Mrs. Piper knew when to end the séance – even though séances lasted from 1 minute to 4.5 hours. He insinuates opposite conclusions to those put forward in *Studies in Spiritism*, simply by saying that "had the book accused her of fakery, Hall and Tanner would have been open to libel suits. If you read carefully between the lines, there are suggestions that in many ways Mrs. Piper practiced deceptions." Gardner also devotes several paragraphs to the negative opinion of William James' sister Alice, even though it has no evidentiary value whatsoever (and given that she refers to Fred Myers as "an idiot", it's difficult to trust her judgement).

33. Amy Tanner, *Studies in Spiritism*, p. 268

34. Ibid. pp.236-238

35. Richard Hodgson, "A Record of Observations of Certain Phenomena of Trance", *Proceedings of the SPR* Volume VIII, p.4

36. Amy Tanner, *Studies in Spiritism*, p. 247

37. Sir Oliver Lodge, "A Record of Observations of Certain Phenomena of Trance", in *Proceedings of the SPR* Volume VI, p. 447.

38. Mrs. Piper correctly named the communicator, the people present at the writing of the letter and where it was done, and paraphrased the writer's reaction if it was communicated by her from the afterlife.

39. Frank Podmore, "Discussion of the Trance-Phenomena of Mrs. Piper", *Proceedings of the SPR* Volume XIV.

40. Ibid.

41. Richard Hodgson, "A Record of Observations of Certain Phenomena of Trance", *Proceedings of the SPR* Volume VIII, p.6

42. William James, 1898, "Letter on Mrs. Piper, the Medium"

43. James Hyslop, "President G. Stanley Hall's and Dr. Amy E. Tanner's Studies in Spiritism", in *Journal of the ASPR* Volume V.

Theo Paijmans - The Fairies' Portal (p. 223)

Notes:

1. Jerome Clark, *The UFO Encyclopedia, the Phenomenon from the Beginning, Vol 1, A-K,* Omnigraphics, 2nd edition, 1998, page 209.

2. 'Lured By Vision To Leave Home, Says Miss Bald. Declares Little Man in Green Urged Her to Make Flight. Denies Rumors Of Any Love Affair', *Philadelphia Inquirer,* Philadelphia, Pennsylvania, 18 Agust 1906.

3. 'Child Thought It Saw Satan. Strange Mystery of Sacred Heart School, Dubuque', *Waterloo Daily Courier,* Waterloo, Iowa, 28 September 1907. Retrieved by Jerome Clark, 11 August 2010.

4. 'Legends of Old Detroit', *Nebraska State Journal,* Lincoln, Nebraska, 25 May 1923.

5. 'Jersey Town Sees Ghost. Weird Apparition Keeps Nervous Residents Home at Night', *The Indianapolis Star,* Indianapolis, Indiana, 19 March 1911.

6. 'A Connecticut Ghost Story', *The New York Times,* New York, New York, 12 January 1886; An Accomodating Ghost. A Massachusetts Dwarf who Appears and Disappears in Night or Day – He is Cut in Two by a Workman's Spade', *St. Louis Globe-Democrat*, St. Louis, Missouri, 12 January 1886.

7. 'Millington's Ghost. A Specter Which is Disturbing a Quiet Village', *St. Louis Globe-Democrat*, St. Louis, Missouri, 13 March 1886; 'Millington's Ghost', *St. Louis Globe-Democrat*, St. Louis, Missouri, 3 April 1886; 'Millington's Ghost. A Strange Supernatural Being Appears to Numerous Persons on Various Occasions – Incredulity Silenced', *Rocky Mountain News,* Denver, Colorado, 11 April 1886.

8. *Proceedings Of The First International UFO Congress,* Curtis G. Fuller eds., Kenneth Arnold, 'How It All Began', Warner Books, 1980, Page 26.

9. 'Saucer, Little Man, Truck, Fog Make Quite a Story', *Evening Sentinel,* Holland, Michigan, 12 October 1954.

10. 'Hunting A Leprechaun. Irish Folk Eager to catch Fairy that's Loose in County Meath', *The Washington Post,* Washington, District of Columbia, 26 April 1908.

11. 'Leprechaun Appears', *Baltimore American*, Baltimore, Maryland, 31 May 1908.

12. 'Police Capture Queer "Leprechaun", *The Washington Post*, Washington, District of Columbia, 16 August 1908.

13. 'Fairies In Ireland. Strange Sights At The Cross-Roads. To The Editor Of The Times', *The Times,* London, United Kingdom, 6 September 1938.

14. Henry-Jean Besset, 'Le Petit Bonhomme de Renève', *Phénomènes Spatiaux,* issue 45, September 1975, pages. 21-27.

15. *Rockport Journal*, Rockport, Indiana, 31 August 1906.

16. 'Here's A Weird Tale. Queer Brand of Liqour Must Be Used in Tennessee', *Suburbanite Economist,* Chicago, Illinois, 28 June, 1907; 'Here's A Weird Tale. Queer Brand of Liqour Must Be Used in Tennessee', *Evening News*, Ada, Oklahoma, 1 July, 1907; 'Here's A Weird Tale. Queer Brand of Liqour Must Be Used in Tennessee', *The Wasatch Wave*, Heber City, Utah, 12 July 1907.

17. Willy Wegner, 'The Mystery Airship Over Denmark In 1908', *UFO Register,* vol.8 no. 1, November, 1977, pages 3-5.

18. Maud Walker, "Aggie and the Tree Fairy", *The Edwardsville Intelligencer,* Edwardsville, Illinois, 18 September 1908.

19. Ethel Holsinger, 'The Moon Fairies' Visit to Mars', *The Washington Post,* Washington, District of Columbia, 9 October 1910.

20. Ruth Hanson, 'On The Green Cheese', *The Oakland Tribune*, Oakland, California, 11 May 1925.

21. 'Daddy's Bedtime Story. The Fairies' Trip to The Stars', *The Edwardsville Intelligencer*, Edwardsville, Illinois, 6 June 1914.

22. Annette Angert, 'On The Yankee Doodle', *Fort Wayne Weekly Sentinel*, Fort Wayne, Indiana, 3 April 1915.

Nigel Watson - Enchanted Islands and Ships in the Sky (p. 243)

References:

Books and Periodicals

Bartholomew, Robert E., and Howard, Professor George S. *UFOs and Alien Contact: Two Centuries of Mystery*, Prometheus Books, 1998.

Bartholomew, Robert E., *Little Green Men, Meowing Nuns and Head-Hunting Panics,* McFarland, Jefferson, 2001.

Bullard, Thomas E., *The Airship File*, privately published, Bloomington, Indiana, 1982.

Bullard, Thomas E., *The Airship File, Supplement I*; *Supplement II*, 1990; *Supplement III,* 1997.

Bullard, Thomas, "Prehistoric UFOs: A Review," *MUFON UFO Journal,* No.243, July 1988, 14-16.

Busby, Michael. *Solving the 1897 Airship Mystery*, Pelican Publishing, 2004.

Fort, Charles. *The Complete Books of Charles Fort,* Dover, New York, 1974.

Ross, Miceal. "Anchors in a three-decker world," *Folklore*, 1998. On the *LookSmart* website at: www.findarticles.com/cf_0/m2386/v109/21250632/print.jhtml

Watson, Nigel (ed.). *The Scareship Mystery,* Domra, 2000.

Watson, Nigel. *The Origin of UFOs: 1807-1917,* e-book at: www.selfhelpguides.com

Winkler, Louis, *Analysis Of UFO-Like Data Before 1947,* Fund For UFO Research, Mount Rainier, USA, 1984.

Websites

Sourcebook Project, at:
www.science-frontiers.com/sourcebk.htm

Tales From The Fort: The Charles Fort Files at:
http://www.dragonrest.net/fortfiles/forthome.html

UFO Historical Review at:
www.cufon.org/uhr/uhrndx.htm

Mr. X, Consulting Resologist. The Fortean Web Site of Mr. X, contains hypertext editions of the works of Charles Fort, at:
http://www.resologist.net/

Notes:

1. Michel Bougard, "UFOs Throughout History," in *UFOs 1947-1987: The 40-Year Search for an Explanation*, ed. Hilary Evans and John Spencer, 20-25 (London: Fortean Tomes,1987).
2. Eric Inglesby, *UFOs and the Christian* (London: Regency Press,1978), 125-133.
3. Erich Von Daniken, *Chariots of the Gods?* (New York: G. P. Putnam's Sons, 1970).
4. Bougard, 20-25.
5. Jacques Vallee, *Magonia: From Folklore to Flying Saucers* (London: Tandem, 1975).
6. Saint Agobard, *De la Grêle et du tonnerre* (Lyon: Imprimerie de Dumoulin, Ronet et Sibuet, 1841). Agobard, Archbishop of Lyons, "De Grandine at Tonitrua," www.chez.com/lesovnis/htm/lyon840.htm (accessed November 7, 2004).
7. Jean-Louis Brodu, "'Magonia: A Re-Evaluation," in *Fortean Studies*. Volume 2, ed. Steve Moore, 198-215 (London: John Brown Publishing, 1995).
8. Ibid., 201-203.
9. Jacques Bergier, *Mysteries of the Earth* (London: Futura, 1973), 146.
10. Vallee, *Magonia: From Folklore to Flying Saucers,* 9-10. John Michell, *The Flying Saucer Vision* (London: Sidgwick & Jackson, 1967), 51-52. John Keel, *UFOs: Operation Trojan Horse* (London: Abacus,1973), 74-75.
11. "Crop Circle News," www.cropcirclenews.com (accessed August 6, 2004).
12. Aubrey Gwynn, ed., *The Writings of Bishop Patrick 1074-1084* (Dublin: Dublin Institute for Advanced Studies, 1955), 65.

13. Wilkins, *Flying Saucers From The Moon*, 156-158.

14. Felix Liebrecht, ed., *Des Gervasius von Tilbury Otia Imperialia* (Hanover: C. Rumpler, 1856), Section 13, 2-3.

15. *The Bible*, King James Version, *1 Thessalonians*, 4:16-17.

16. Daniel Cohen, *The Great Airship Mystery* (New York: Dodd, Mead & Company, 1981), 79.

17. Robert G. Neeley, *UFOs of 1896/1897: The Airship Wave* (Mount Rainier: Fund For UFO Research, 1988), 66-67.

18. Whitley Strieber, *Communion: A True Story* (London: Arrow Books, 1988), 240.

19. Neeley, 189.

20. Cohen, 79.

21. Keel, *UFOs: Operation Trojan Horse*, 74-75 and 171-174.

22. "Air Ship Takes A Cow. The Thrilling Experience of A Woodson Stockman," *The Farmer's Advocate* (Yates Center, Kansas), April 23, 1897.

23. Cohen, 98.

24. Thomas Bullard, *The Airship File* (Bloomington: privately published, 1982), 104.

25. Jerome Clark, "The Great Airship Hoax," *Fate*, February, 1977.

26. Nigel Watson, "Down to Earth," *Magonia*, No. 44, October 1992, at: http://magonia.haaan.com/2009/down-to-earth/

27. Cohen, 52-153.

Lightning Source UK Ltd.
Milton Keynes UK
UKOW04f1513191114

241866UK00001B/175/P

9 780980 711141